Primary Math Puzzlers

by Ann Fisher

illustrated by Rob Funderburk

Cover by Amy Hesselbacher

Copyright © 1994, Good Apple

ISBN No. 0-86653-812-7

Printing No. 9876543

Good Apple
1204 Buchanan St., Box 299
Carthage, IL 62321-0299

Paramount Publishing

With special thanks to . . .

Kim, Phil, Jackie and Greg
and to my loving family

Table of Contents

ONE!
TWO!
THREE!

GA1504

To the Teacher

Primary Math Puzzlers is a book packed with a fresh variety of puzzles and activities that will appeal to your students and reinforce valuable skills in mathematics. Its pages can provide enrichment work for independent students and five-minute fill-ins for the entire class. It is the author's hope that through these activities, students will find the joy of "playing" with numbers and mathematical concepts.

Most pages in this book will fall into one of these categories:
1. Puzzles that provide fun computational practice or
2. Puzzles that emphasize problem-solving skills.

For younger students there are many pages with cutout manipulatives. This allows them to try out various solutions without the frustration of erasing. For older students there are pages to challenge them to find multiple solutions and to apply acquired skills to new situations.

The puzzles have been arranged into sections by broad topics, as listed in the Table of Contents. Easier puzzles are usually placed first within each section. Since many puzzles cover overlapping skills, you may want to skim other parts of the book when looking for a puzzle to cover a specific skill. For example, some puzzles in the "Money and Measurement" pages require skills in addition. To help you, specific skills used on each page have been listed in the top right-hand corner.

Carefully preview each puzzle before presenting it to your students to be sure it is appropriate for their skill level. And especially with younger students, go over the instructions for each page with the students before assigning independent work.

As students' abilities and interests develop, encourage them to write their own math puzzles for others (maybe even *you*) to solve. Happy puzzling!

Buying Buttons

Nancy wants to buy exactly 16 buttons to sew on her new coat.

1. Circle the card of buttons that she should buy.

A.
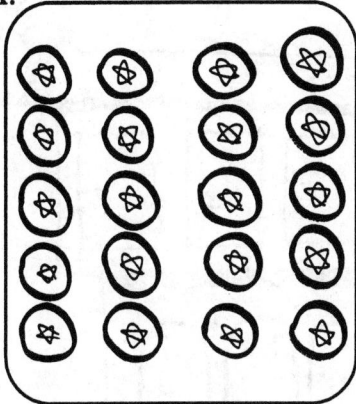

B.
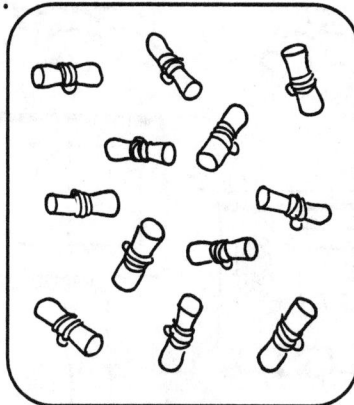

C.

D.

E.

F.
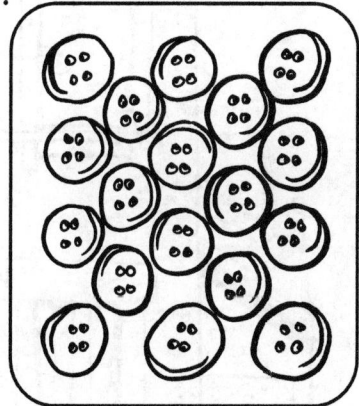

2. Now go back and look at the cards again. On the ones that have *less* than 16, add enough buttons to make 16.

3. On the cards that have *more* than 16, cross out the extra buttons.

1

King Key

King Key likes to carry lots of keys. In fact, he has a separate key for every door on his castle. Circle the number below that shows how many keys he will need to unlock every door in the castle.

25   27   29

Now draw all of the king's keys on the back of this page.

**Use two colors of crayons to color the castle doors in some kind of pattern.

GA1504

Locker Duty

You are in charge of painting numbers on some of the lockers at your school. You will number them from 1 to 40. How many times will you have to paint the

number 2? _____

Leap Frog

Frank and Franny are ready to play in their favorite pond. They will jump along the paths shown here.

NO SWIMMING

a. First, number the spaces on each frog's path.

b. Then cut out the frog markers at the top of the page.

c. Next, move Frank along his path. With each leap, Frank moves 2 spaces.

 How many jumps will it take Frank to get into the pond? _____

d. Now move Franny along her path. She moves 3 spaces with each leap. How

 many jumps will it take Franny to get into the pond? _____

4

GA1504

Three's the Key

Start at the ③ and connect the numbers in order that will help you count by 3's to ㉝. You will draw a familiar shape.

100	86	85	86	87	91	94	99	100
98	82	83	84	89	90	92	95	97
80	79	81	79	16	14	㊼93	20	19
77	78—③	6	9	12	15	18	22
76	75	5	8	13	11	19	21	20
68	72	70	52	49	47	26	24	26
65	69	67	51	48	45	46	27	31
64	66	61	54	50	42	41	30	34
62	63	60	57	58	39	36	33	40

GA1504

Sort It!

This machine sorts odd numbers and even numbers. Show where each number will land by writing it in the correct bin.

| 13 | 25 | 4 | 7 | 14 | 9 | 6 | 12 | 10 | 3 | 8 | 1 |

IN

OUT

ODD

EVEN

Odd   **Even**

GA1504

Counting Scramble

First unscramble these number words and write each word in the blank. (Use the word list for help.) Then find a place in the puzzle tree where each word will fit.

inen _____   urof _____

githe _____   ixs _____

nosedc _____   trufoh _____

ent _____   vief _____

neves _____   weyntt _____

neo _____   nehirtte _____

drith _____

Word Bank

one	fourteen
two	fifteen
three	sixteen
four	seventeen
five	eighteen
six	nineteen
seven	twenty
eight	first
nine	second
ten	third
eleven	fourth
twelve	fifth
thirteen	

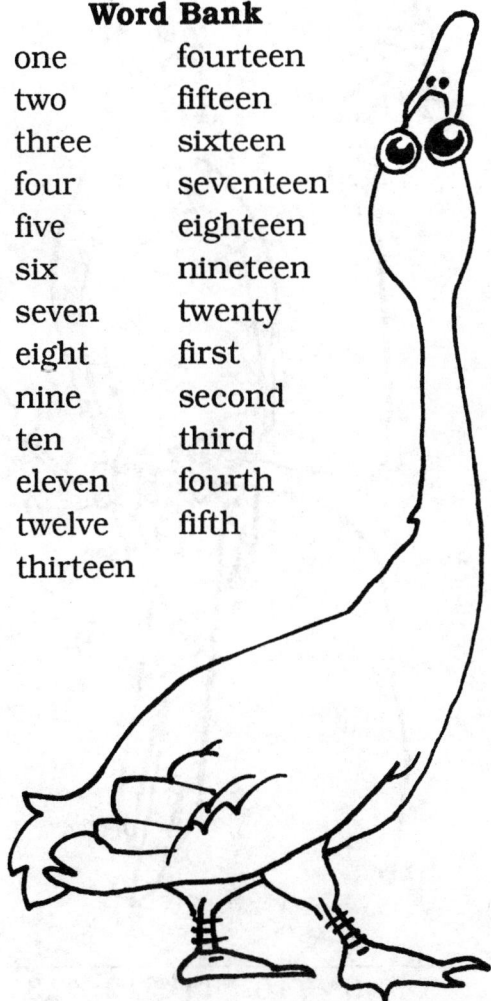

```
__ __ C __ __ __
   __ O __ __
__ __ U __ __ __
   __ N __
   T __ __
 __ I __ __
__ __ __ __ __ N
 __ __ G __ __
  __ W __ __ __ __
     O __ __
 __ __ __ R __ __ __ __
__ __ __ __ D __ __
        S __ __
```

Animal Antics

What do you get if you cross a centipede with a parrot? To learn the answer, find each letter given in the directions below the chart. Then write those letters (in order) at the bottom of the page.

A B C D E F G H I J
K L M N O P Q R S T
U V W X Y Z A B C D
E F G H I J K L M N

1. first letter in the first row _____

2. third letter in the third row _____

3. seventh letter in the third row _____

4. second letter in the second row _____

5. seventh letter in the fourth row _____

6. ninth letter in the first row _____

7. fifth letter in the first row _____

8. tenth letter in the second row _____

9. seventh letter in the third row _____

10. eighth letter in the fourth row _____

11. first letter in the second row _____

12. fifth letter in the fourth row _____

13. first letter in the fourth row _____

Answer: ___ ___ ___ ___ ___ ___ ___ – ___ ___ ___ ___ ___ ___
        1   2 3 4 5 6 7   8 9 10 11 12 13

GA1504

Ann's Antiques

Ann collects all kinds of antiques. Here are a few interesting items from her collection. She's found the year each one was made, and now she wants to arrange them in order from the oldest to the newest. Can you help her? Put a 1 in the blank under the oldest one, a 2 under the second oldest, and so on.

thimble
1684

_____

old-fashioned
electric toaster
1930

_____

"Brownie"
camera
1902

_____

safety
pin
1851

_____

saxophone
1845

_____

scissors
1775

_____

upright
piano
1911

_____

microwave
oven
1962

_____

eye-
glasses
1543

_____

stapler
1876

_____

GA1504

What Fits?

Look at each group of numbers below. Then read all of the labels. Draw a line from each box to the label that describes *every* number in that box. (There will be two extra tags.)

Box A

26
22
18
4
14

Box B

11
13
7
3
9

Box C

9
8
5
6
2

Numbers with 1 ten

Numbers less than 10

Odd numbers

Even numbers

Numbers greater than 20

GA1504

Number Detective

Your job is to find a place in this chart for each number below. First cut out the number pieces and move them around. When you have found a spot for every number, write the numbers in the places where they belong. (Some numbers look like they can go in more than one spot, but you need to find a solution where every number has a place.)

Numbers with less than 5 ones			
Numbers with 2 tens			
Numbers greater than 30			

31	25	14	21	36	23	34	28	26

GA1504

Teamwork! (Addition)

3 + 1 = 4 is a number sentence.

1 + 3 = 4 is another number sentence that belongs on the same team because it shows the same fact in a different way.

Now you can help some other number sentences team up. First, write the answer to each one. Then cut out all the sentences, find a teammate for each, and glue each team into the box that shows the right answer for that team.

	5	**6**
	7	**8**

2 + 3 = ___

3 + 2 = ___

4 + 2 = ___

2 + 4 = ___

3 + 5 = ___

5 + 3 = ___

4 + 3 = ___

3 + 4 = ___

GA1504

Teamwork! (Subtraction)

Now let's put some subtraction facts on the teams. For the number sentences
3 + 1 = 4 and 1 + 3 = 4 there are two subtraction facts that can join the team.
They are 4 - 1 = 3 and 4 - 3 = 1.

Now write the answer to each number sentence below. Then decide which team
each should join and glue it there.

3 + 2 = 5 2 + 3 = 5	2 + 4 = 6 4 + 2 = 6
4 + 3 = 7 3 + 4 = 7	5 + 3 = 8 3 + 5 = 8

5 - 2 = ___   6 - 2 = ___

8 - 3 = ___   8 - 5 = ___

6 - 4 = ___   5 - 3 = ___

7 - 3 = ___   7 - 4 = ___

GA1504

Domino Dots I

Gary, Amy, and Lee are drawing tiles from their domino set. This chart shows what each player draws. For each turn, circle the domino with the highest total and put an *X* on the domino with the lowest total.

First Turn   Gary   Amy   Lee

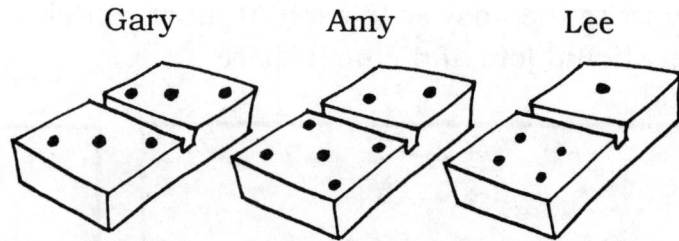

Second Turn

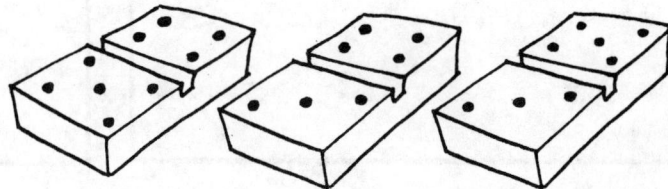

Third Turn

Fourth Turn–Draw dots on the last three tiles so that Amy has the highest total and Lee has the lowest total.

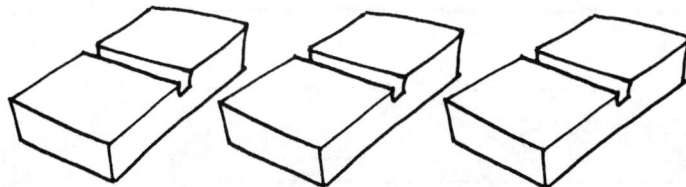

Game Idea

Get a pile of dominoes and a friend. Take turns drawing tiles and write down each person's total score. After ten turns each, see who scored the highest total most often.

Domino Dots II

Emily is playing a domino game with her sister, Maddie, and wants to find a tile with a total of 7 dots on her next turn. Draw dots on these dominoes to show 3 different ways Emily can score a total of 7.

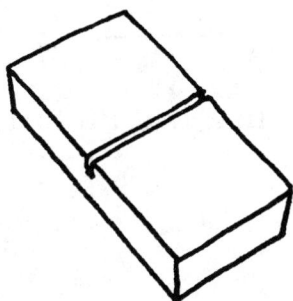

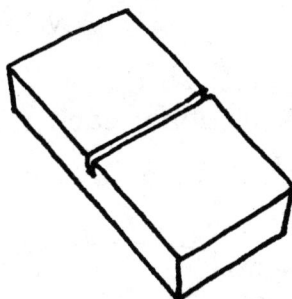

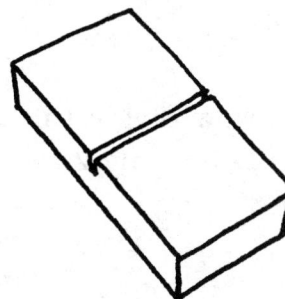

Now Maddie wants to find a total of 8 dots on a tile. Draw dots to show 3 different ways she can score 8.

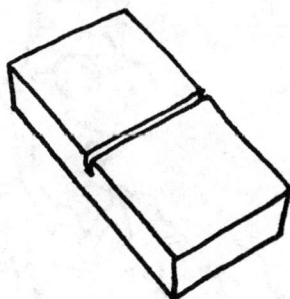

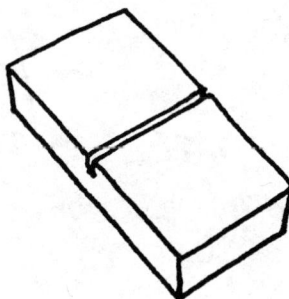

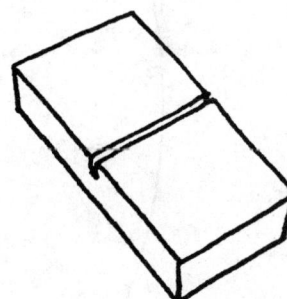

Circular Sums

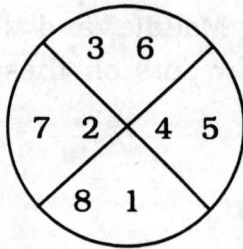

This circle has been divided into 4 sections by drawing 2 intersecting lines. When you add together the numbers in each section of the circle, you always get the same sum.

What is it? _____

Now divide this circle into 4 sections, each with numbers that add up to 15. Draw only 2 intersecting lines.

16

GA1504

Problem Search

You've heard of a word search puzzle before, and now here's a math problem search puzzle for you. The problems have already been solved–all you have to do is find them! For example, in the top row, we've circled the problem 3 + 6 = 9. (We had to put in the + and = signs.) You should be able to find many more problems inside this puzzle. They will read either from left to right or from top to bottom. Add + or - and = signs as needed. Circle each problem that you find and write each one below too. Numbers in the puzzle may be used more than once.

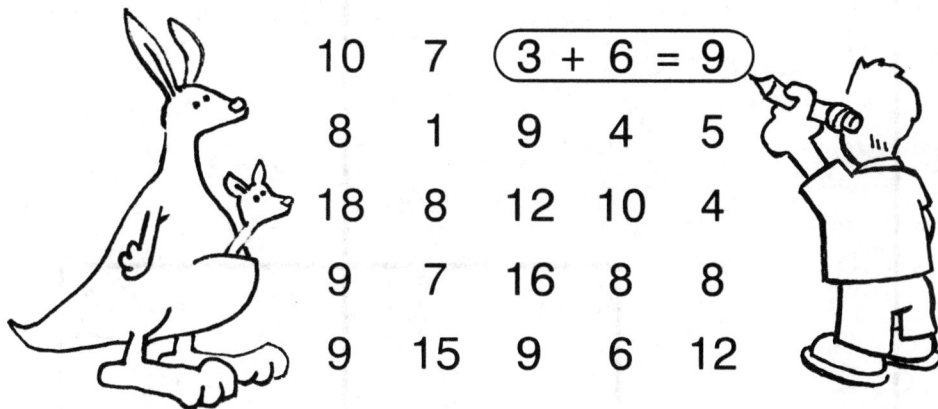

```
10   7   (3 + 6 = 9)
 8   1   9   4   5
18   8  12  10   4
 9   7  16   8   8
 9  15   9   6  12
```

Across

a. 3 + 6 = 9

b. _____

c. _____

d. _____

e. _____

f. _____

g. _____

Down

a. _____

b. _____

c. _____

d. _____

e. _____

f. _____

g. _____

h. _____

Sum Squares

Can you arrange the numbers 0, 1, 2, 3, 4, 5, 6, 7, 8, and 9 so that the four corners of each of the three squares add up to 19? (Cut out the numbered markers below and move them around. When you've found a solution, write the numbers in place and remove the markers. This will save you a lot of erasing!) The 3 has been placed to help you get started.

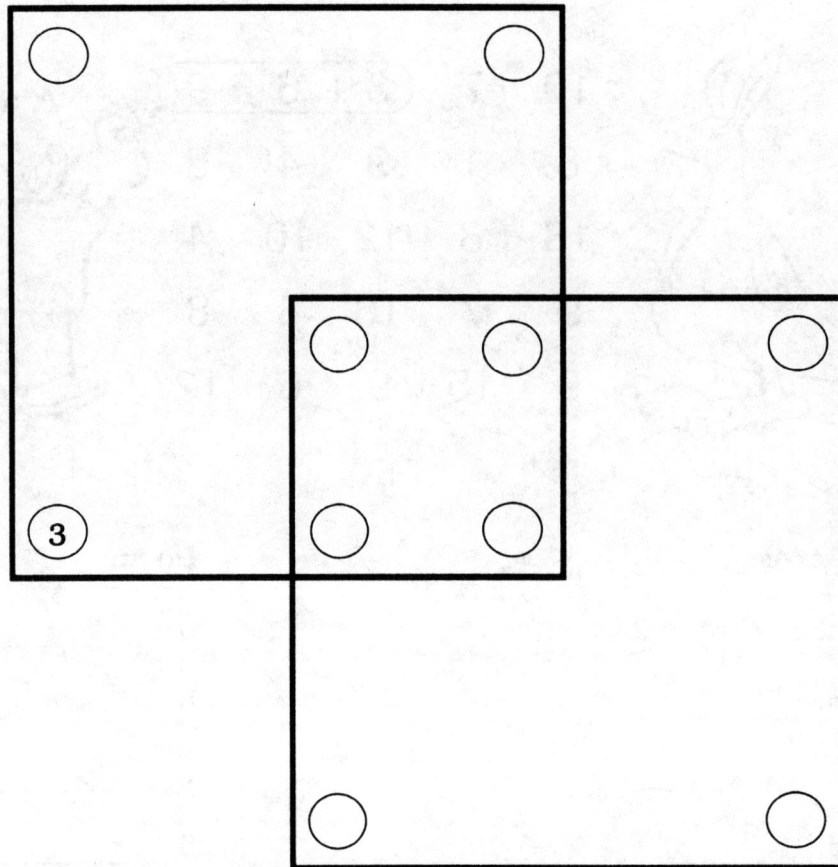

| 0 | 1 | 2 | 3 | 4 | 5 | 6 | 7 | 8 | 9 |

GA1504

Pick-a-Path

1. Pick a path starting at A and ending at B. Add the numbers in the circles together as you pass through them. Try to find the path with the highest total. You must follow the lines drawn on the path, and you may enter each circle only once.

   Numbers on your path: _____

   Total: _____

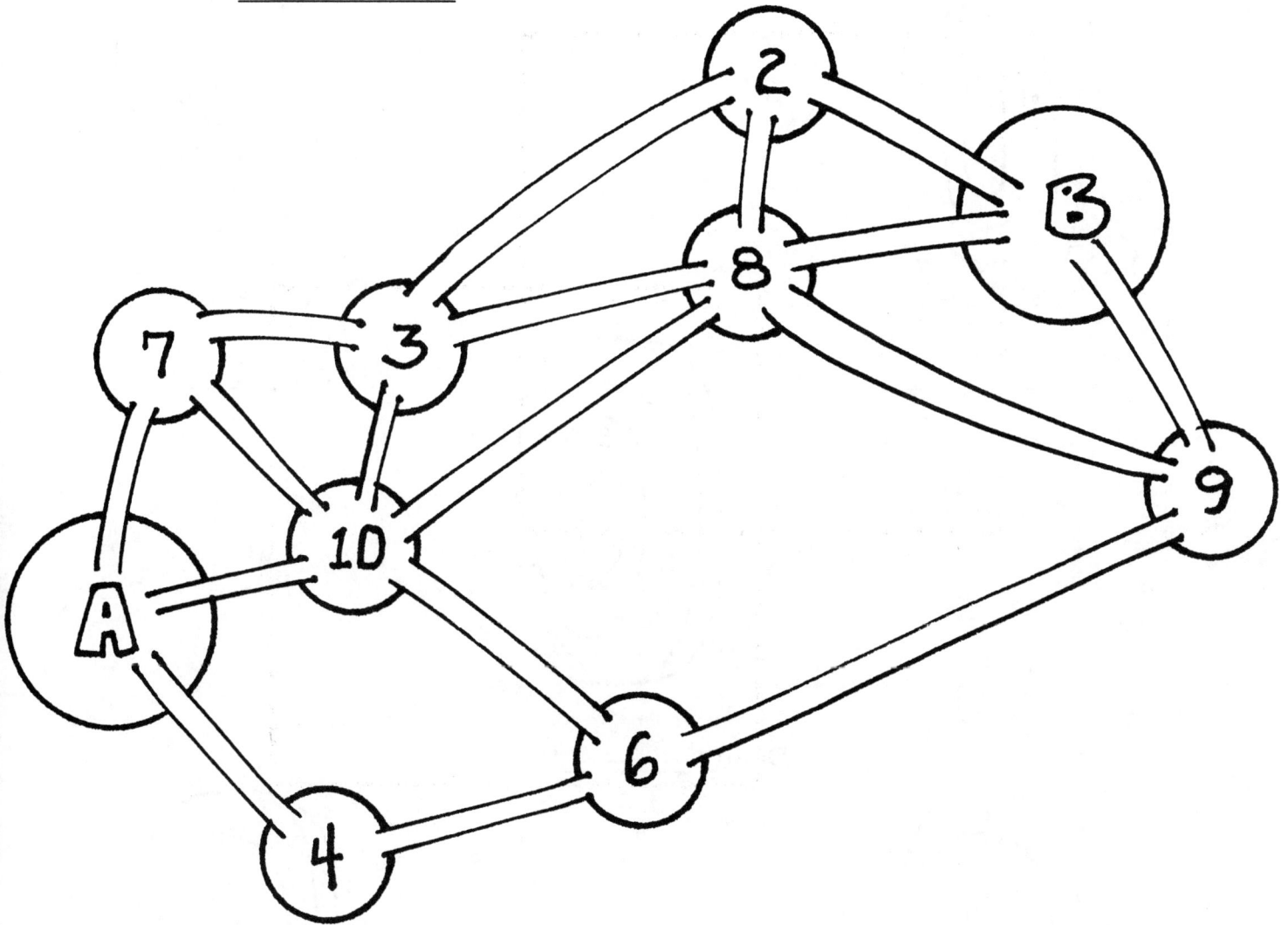

2. Now pick the path with the lowest total.

   Numbers on your path: _____

   Total: _____

Bull's-Eye I

Five friends have been practicing their archery. Each person's target is shown here. The dots tell where each arrow landed. Add the value of each area of the target that has been hit. Write each person's score below his or her target. Circle the name of the person with the highest score.

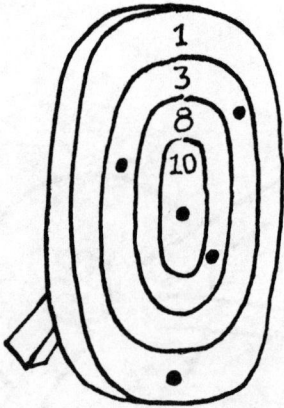

Greg _____

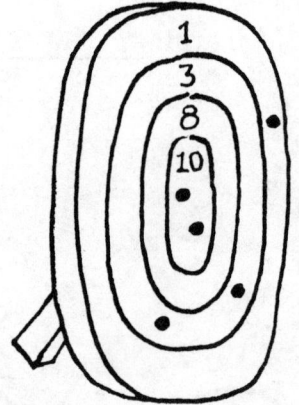

Jan _____

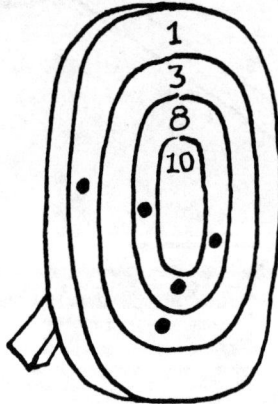

Danny _____

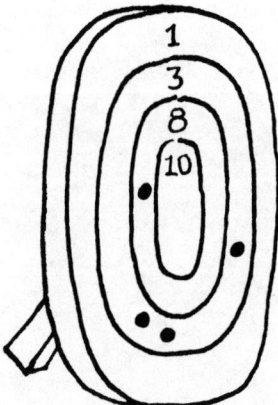

Kate _____

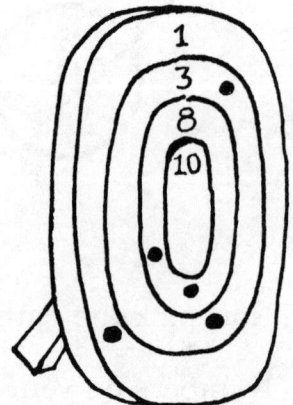

David _____

GA1504

Bull's-Eye II

Greg, Jan, Danny, Kate, and David are practicing archery again. Here are the targets they are using. Put dots on each to show one way each player might have scored the points shown. Each player shot five arrows.

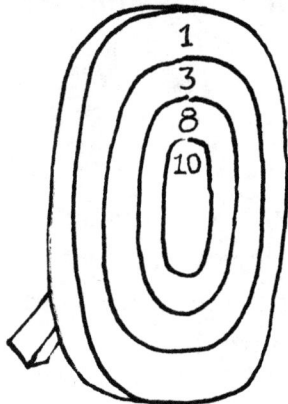

Greg–32

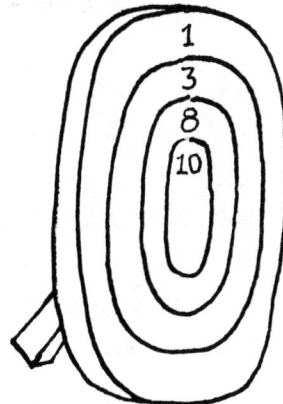

Jan–35

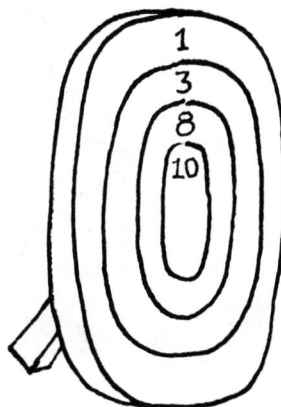

Danny–30

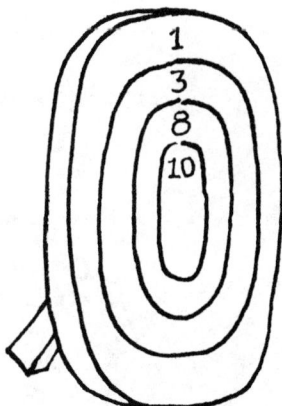

Kate–39

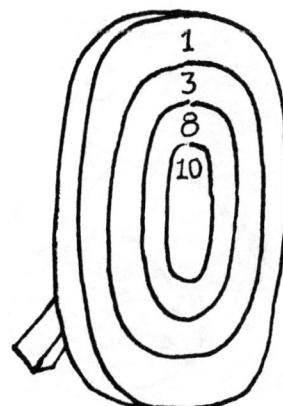

David–44

**Write down a number that could have been your score. See if a classmate can show where your arrows went.

21

GA1504

Pet Store Prices

A = 1¢
B = 2¢
C = 3¢
D = 4¢
E = 5¢
F = 6¢
G = 7¢
H = 8¢
I = 9¢
J = 10¢
K = 11¢
L = 12¢
M = 13¢
N = 14¢
O = 15¢
P = 16¢
Q = 17¢
R = 18¢
S = 19¢
T = 20¢
U = 21¢
V = 22¢
W = 23¢
X = 24¢
Y = 25¢
Z = 26¢

Pete's Pet Store is having an unusual sale. Each animal is priced according to the letters in its name. Each letter is priced as shown in the chart. To find the cost of a DOG, for example, you would add:

 4¢ + 15¢ + 7¢ for a total of 26¢.
 (D)   (O)   (G)

Now look at the animals shown here and on the next page. First circle the one you think will have the lowest price and underline the one you think will have the highest price. Then find the actual price for each and write it under each animal.

Mouse	Rabbit	Kitten

_____	_____	_____

Crab	Snake	Lizard

_____	_____	_____

GA1504

Pet Store Prices (cont'd.)

Puppy   Iguana   Parrot

_____   _____   _____

Now, try to think of a pet Pete's store might be selling that costs more than $1.00.

Animal: _____   Price: _____

****

Work Space:

GA1504

Domino Dilemma

Jim and Barb are playing dominoes. The set they are using goes from double blanks up to double sixes. All of the tiles have been placed facedown, and each player selects six.

1. Barb discovers that she has drawn the lowest number of dots possible for six dominoes. One domino she drew is, of course, the 0 + 0 (or double blank). Another is the 0 + 1. What are the other four tiles Barb has?

   _____ _____ _____ _____

2. How many dots does Barb have altogether? _____

3. Jim learns that he has the highest number of dots possible on his six tiles. What six dominoes did he get?

   _____ _____ _____

   _____ _____ _____

4. How many dots does Jim have? _____

5. Now suppose that Jim and Barb are playing with a set of dominoes that goes up to double nines. If Jim again drew the six tiles with the highest number

   of dots, how many dots would he have? _____

6. If Jim drew the highest six tiles from a set of double twelve dominoes, how

   many dots would he have? _____

GA1504

Book Drops

Three library vans each started the day with 100 books to deliver to other branches of their library. At each stop they dropped off the number of books shown in the circles. First look at all three paths. Which van do you think will

have the most books left at the end of the day? _____

Which van will have the fewest left? _____

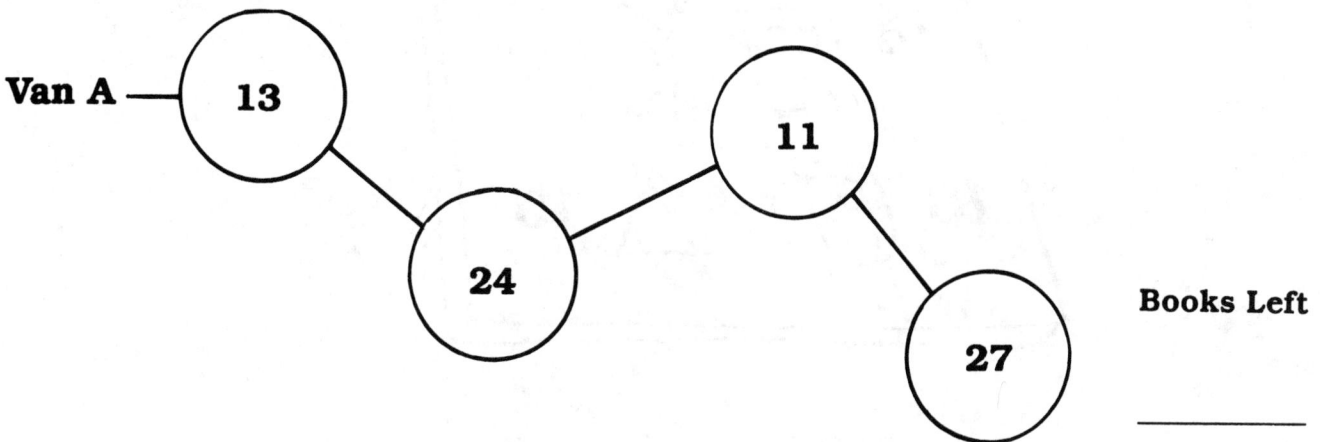

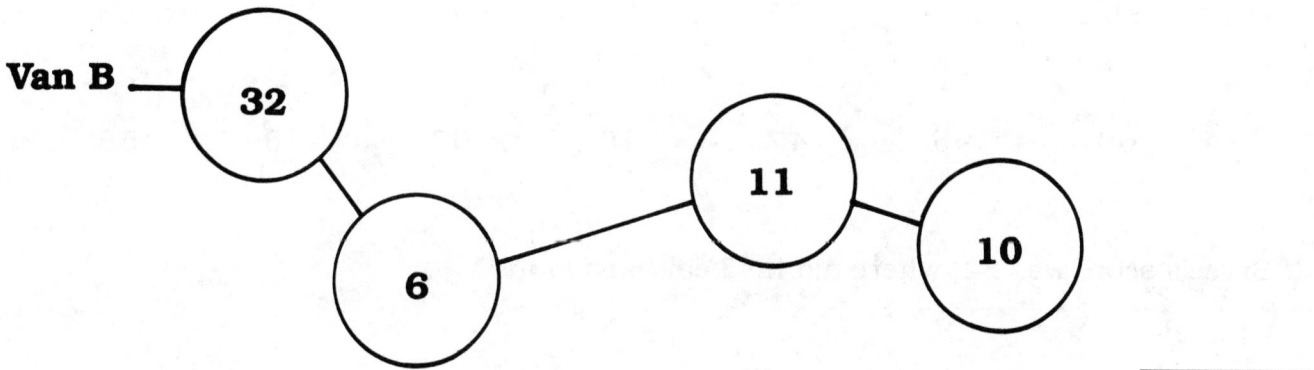

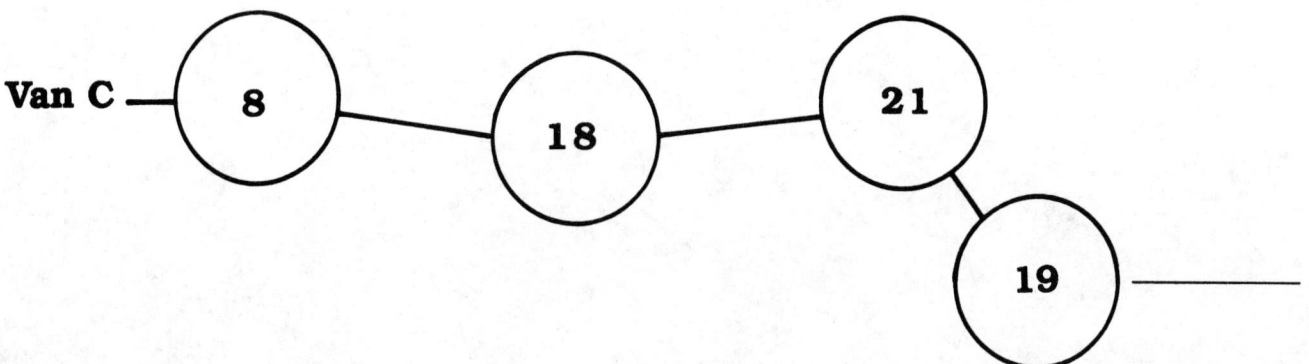

Van A

13   24   11   27   **Books Left** _____

Van B

32   6   11   10   _____

Van C

8   18   21   19   _____

Beanbags

Bryce threw 4 beanbags. All of the beanbags went through holes in this board. (More than one bag may have gone through any of the holes.) Bryce added the numbers above the holes where his bags landed to find his score. Circle the numbers here that could have been his score.

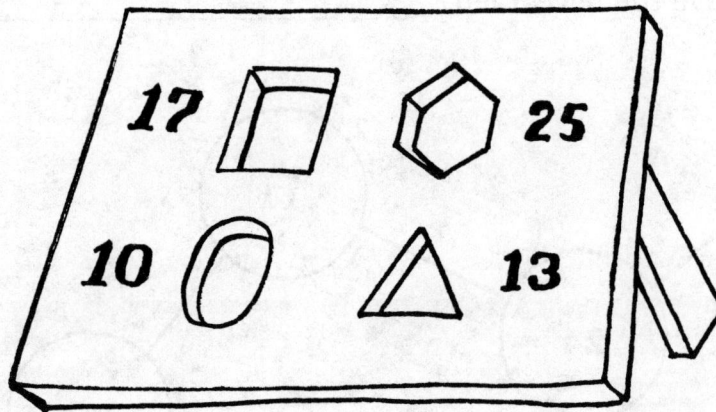

| 65 | 80 | 100 | 75 | 49 | 60 | 107 | 85 |

| 39 | 68 | 95 | 47 | 46 | 62 | 50 | 56 |

If Bryce's score was 84, where did his beanbags land?

_____

GA1504

Bingo!

Can you supply the missing numbers on this BINGO card? Use the information already found on the card and the three clues below. (The numbers under each letter indicate which numbers may appear in that column.)

B (1-15)	I (16-30)	N (31-45)	G (46-60)	O (61-75)
11	Ⓑ	37	50	69
Ⓐ	17	41	47	63
1	29	F R E E	53	70
7	26	44	Ⓒ	75
3	23	35	54	Ⓓ

1. The two diagonal bingos have the same sum. The four corners also share that sum.

2. The missing number in the I column is the lowest in that column.

3. The second horizontal row and the bottom horizontal row have the same sum.

Test Drive

Five drivers each test-drove a new car. The chart shows their beginning odometer reading and their ending reading. By subtracting the beginning mileage from the ending mileage, you will find the number of miles each person drove.

First circle the name of the driver that you think drove the most miles. Then do the subtraction and write the actual number of miles each one drove in the spaces in the chart.

	Starting Mileage	Ending Mileage	Total Miles
Clint	478	1147	
Michelle	364	1260	
Ken	1003	1590	
Elizabeth	1136	1628	
Robert	691	1663	

GA1504

Test Drive (cont'd.)

From your subtraction work, you should have found one driver who drove 587 miles. That driver's name was _____. Study this map. If this driver started in one city, drove directly to a second city, and then directly to a third city, there are two routes he could have taken that would have resulted in a trip of exactly 587 miles. Find the routes on the map and then fill in the blanks below.

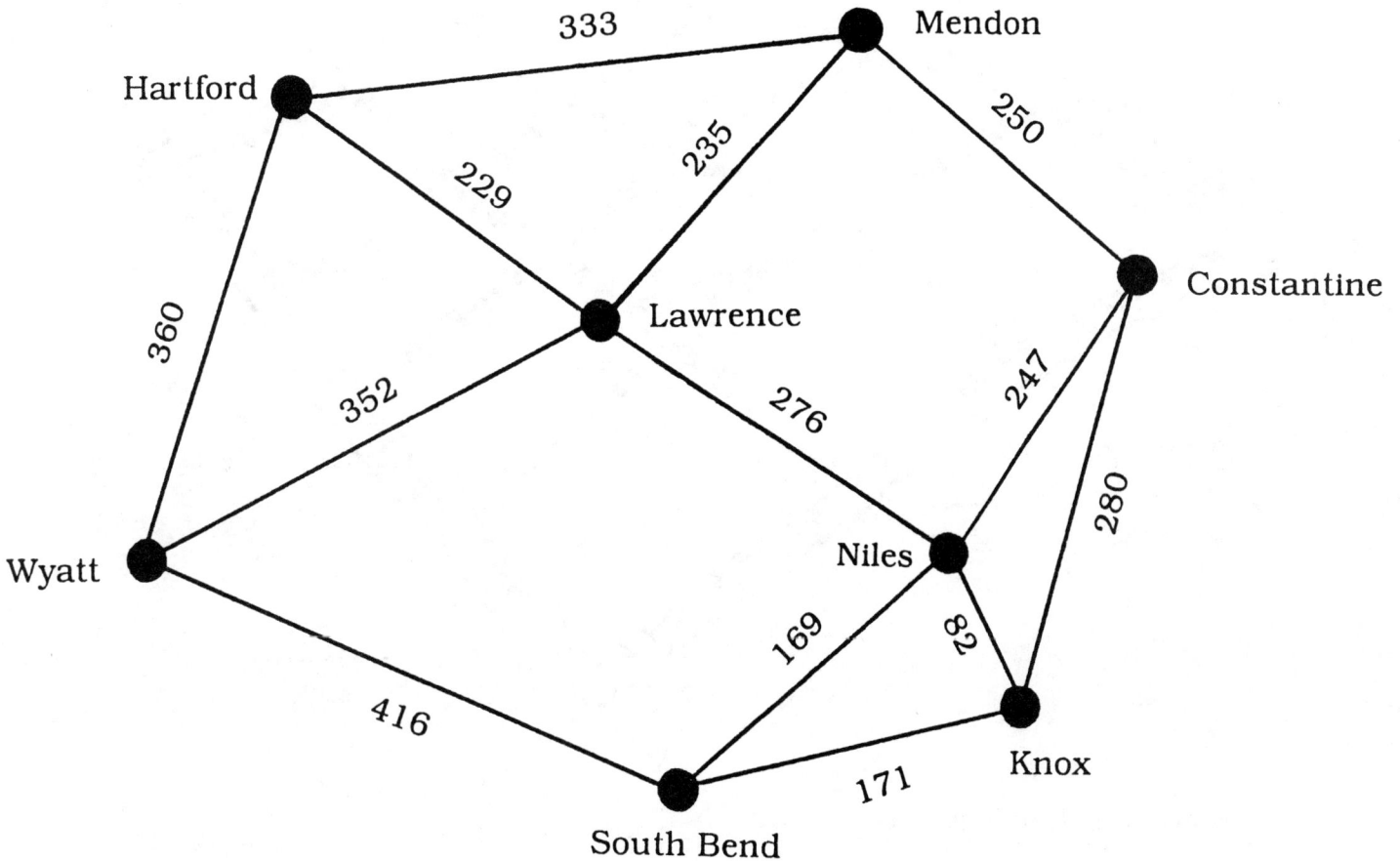

Route #1: Start in _____, drive to _____, end in _____.

Route #2: Start in _____, drive to _____, end in _____.

**Can you find a third route that totals 587 miles and takes you into a fourth city? Start in _____, go to _____, then to _____, and end in _____.

29

Quilt Crafts

Three friends, Lindsey, Carrie and Avery, are going to work together to make this quilt.

1. How many quilt squares are there in all? _____

2. If the three friends share the work, how many squares will each one make?

   _____

3. Carrie decides to decorate each of her squares with 2 flowers. How many

   flowers will she need in all? _____

4. Invent your own quilt pattern and decorate the quilt.

Carnival Candy

Mark and Sally went to the carnival, and both of them bought candy.

This is what Mark bought:

A. _____ pieces for 2¢ each

_____ ¢ in all

This is what Sally bought:

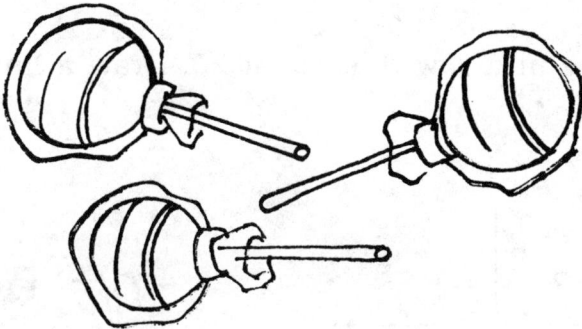

B. _____ pieces for 3¢ each

_____ ¢ in all

C. Who spent more? _____

D. **If Sally had spent 15¢, how many pieces would she have bought?

_____

GA1504

Tic-Tac-Twelve

In this tic-tac-toe board find 3 problems in a row that equal 12. Draw a line through that row.

A.

3 x 4	5 x 3	4 x 4
2 x 2 x 3	2 x 1 x 3	2 x 4
3 x 2 x 2	4 x 3	5 x 2

In this tic-tac-toe board, find 3 problems in a row that equal 20. Draw a line through that row.

B.

3 x 1 x 3	3 x 5	4 x 5
5 x 5	3 x 3	5 x 2
4 x 5	2 x 2 x 5	5 x 4

Potato Problem

First find the answer to each division problem. Then look at the chart to find out what letter each number represents. Put the matching letters in the bottom boxes. If you've done each problem correctly, you'll find the answer to this division riddle.

Riddle:
If you had 5 potatoes and had to divide them equally among 3 people, what should you do?

18 ÷ 2	10 ÷ 5	36 ÷ 6	16 ÷ 4

25 ÷ 5	28 ÷ 7	15 ÷ 5	27 ÷ 3

32 ÷ 4	21 ÷ 3	9 ÷ 9	42 ÷ 7	30 ÷ 6

1 = R

2 = A

3 = E

4 = H

5 = T

6 = S

7 = I

8 = F

9 = M

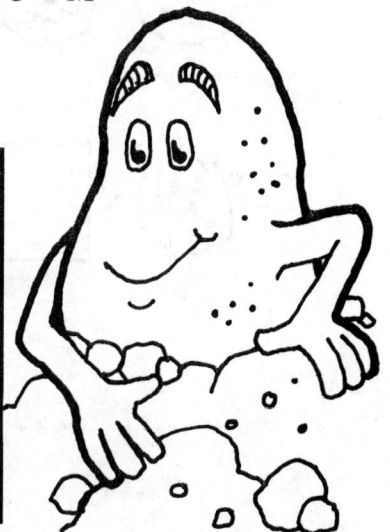

Jack-O-Math

Each jack-o'-lantern face stands for a number. For example, ⬭ is 0, 🎃 is 2, 🎃 is 6, and 🎃 is 8. The faces can be used in the ones place or the tens place. For example, 🎃 🎃 is 28.

Try to figure out the values of the other numbers and solve each problem. Write your answers in "jack-o-math" pictures in the boxes.

1. 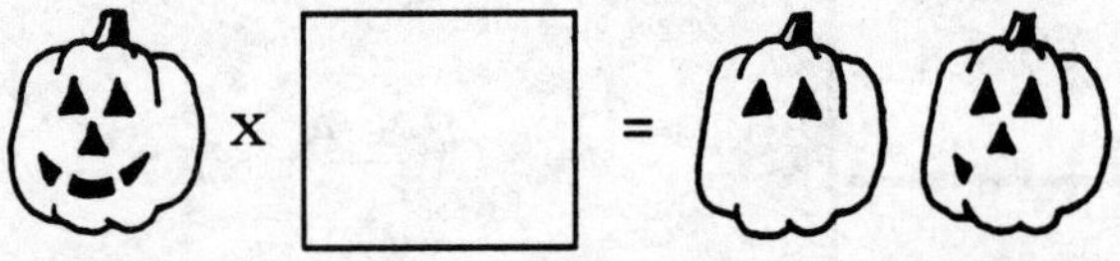 🎃 x [   ] = 🎃 🎃

2.  🎃 🎃 x 🎃 = [   ] [   ]

3.  🎃 🎃 ÷ 🎃 = [   ]

4. 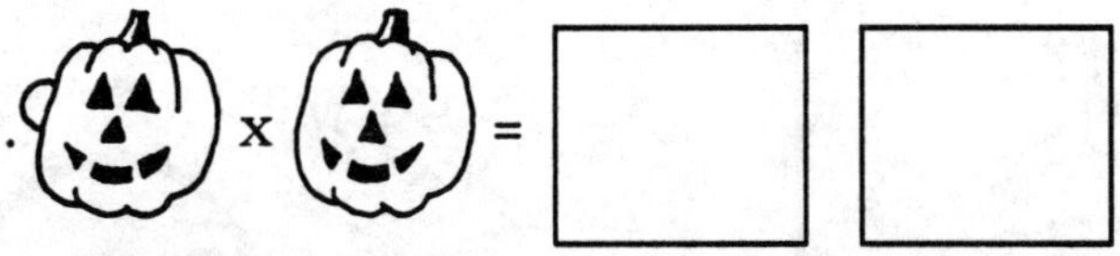 🎃 x 🎃 = [   ] [   ]

5. 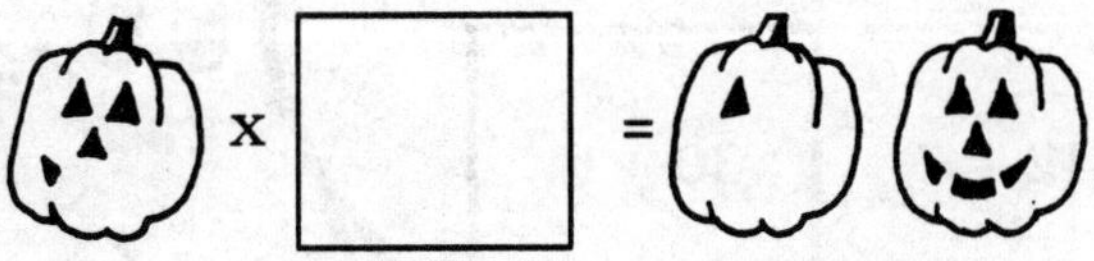 🎃 x [   ] = 🎃 🎃

GA1504

Dice Doodles

Imagine that each cube below is a die with numbers from 1 to 6. Also imagine that you roll each pair of dice and then multiply the numbers that turn up. For example, if you roll a 2 and a 3, your answer would be 6, because 2 x 3 = 6.

Now draw dots on each pair of dice to show all the ways you could get answers that are equal to or greater than 10. We've done one to help get you started. (Show each multiplication fact only once–2 x 5 is the same as 5 x 2.)

Example:

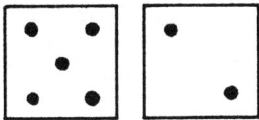

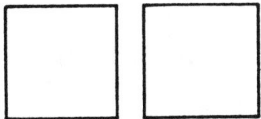   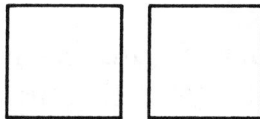   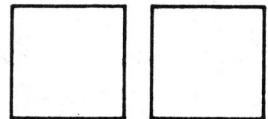

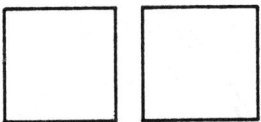   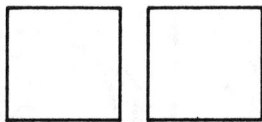   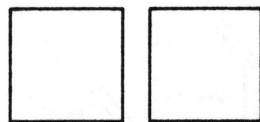   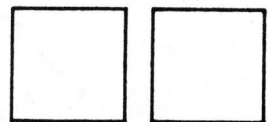

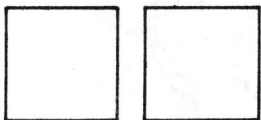   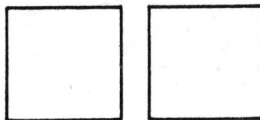   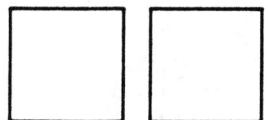

GA1504

A-Maze-ing 40

Find a path through this maze of numbers where the product of the numbers on each side of the line is always greater than 40. We've started the line between the 6 and 7 because 6 x 7 = 42, and 42 is greater than 40. Next we move to 7 - 7 because 7 x 7 = 49 which is also greater than 40. Continue the line until you get "out" of the maze on the right-hand side.

6	5	0	6	1	3	4	9	9
7	7	4	4	2	0	7	7	5
3	8	8	5	1	8	6	2	0
4	3	6	9	9	7	8	5	2
5	0	1	3	6	6	1	4	10
6	3	7	4	5	9	5	7	3

There are two 3 x 2 arrangements of numbers in the maze where all 6 numbers are factors of 40. Find them and draw a box around each.

36

GA1504

Cross Out

Follow the directions below to cross out many of the numbers in this chart. (Some numbers will be crossed out more than once.) When you're done, unscramble the letters next to the remaining numbers. Write the word you spell in the blanks at the bottom of the page to finish a message.

C 6	O 45	J 3	A 10	G 26
R 13	X 20	F 22	M 4	P 14
Q 49	E 2	B 21	Y 42	K 15
W 40	L 28	D 5	Z 35	E 17
H 12	A 16	V 7	S 24	N 11
U 1	N 30	T 9	I 50	L 25

1. Cross out all factors of 24.
2. Cross out all multiples of 5.
3. Cross out all multiples of 7.
4. Cross out all factors of 22.

Message: You're __ __ __ __ __ !

Missing Signs

Place +, -, x, ÷, and = signs in the number sentences below to make each one true. Remember to work from left to right. Sometimes you will need to add ( ) to show which step comes first.

Examples: 3 6 18   3 x 6 = 18

            3 15 5   3 = 15 ÷ 5

            4 3 2 = 10   (4 x 3) - 2 = 10

1. 4   4   16

2. 3   21   7

3. 20   10   2

4. 8   3   24

5. 5   4   2   =   2

6. 6   3   5   =   10

7. 12   3   4   =   8

8. 10   4   6   =   12

9. 16   8   2   =   0

10. 15   10   5   =   20

GA1504

Bingo 48

Find 5 squares in a row where all the answers equal 48. Your row may go horizontally, vertically, or diagonally.

4 x 4 x 3	2 x 24	2 x 5 x 5	48 ÷ 1	192 ÷ 4
2 x 2 x 12	126 ÷ 3	3 x 4 x 4	7 x 7	3 x 2 x 2 x 4
3 x 4 x 5	2 x 2 x 2 x 6	2 x 4 x 6	96 ÷ 2	48 ÷ 1
3 x 16	96 ÷ 2	2 x 3 x 8	4 x 12	144 ÷ 3
144 ÷ 3	192 ÷ 4	24 x 2	16 x 3	72 ÷ 2

THERE WAS A FARMER - HAD A DOG - AND BINGO WAS HIS NAME - O...

GA1504

What's Left?

This is a puzzle about remainders–the "leftover" numbers after you divide one number by another. Your job is to find a "home" in this chart for *every* number below. The number 13 has been placed in the first box because when you divide it by 3, your answer is 4 with a *remainder of 1*. Now cut out the other numbers below and move them around in the chart until every number has a spot. When you've found a solution, write the numbers into the boxes and remove the markers.

	Remainder 1	Remainder 2	Remainder 3	Remainder 4	Remainder 5
Divide by 3	13		✗	✗	✗
Divide by 4				✗	✗
Divide by 5					✗
Divide by 6					

**Why are some of the boxes in the chart crossed out?

7	8	15	17	18
21	22	23	26	28

34	41	49

40

Disappearing Digits

These division problems *were* finished, but now some of the numbers have disappeared. Can you figure them out? Write the correct number in each empty box.

```
        3 □ 9
   6 │ 1 9 7 □
       1 □
       ─────
       1 7
       1 2
       ─────
           5 □
           5 □
       ─────────
             0
```

```
        □ 1 5 □
   7 │ 4 3 □ 6 4
       4 2
       ─────
       □ 0
         7
       ─────
         □ 6
         3 □
       ─────
           □ 4
           1 4
         ─────
             0
```

```
        8 4 □ 1
   □ │ 6 7 8 4 8
       6 4
       ─────
       □ 8
       3 2
       ─────
         □ 4
         6 □
       ─────
           0 8
             8
         ─────
             0
```

```
        □ 1 □ □
   9 │ 7 3 □ 4 2
       □ □
       ─────
       1 2
         9
       ─────
         □ 4
         2 7
       ─────
           □ 2
          -7 2
         ─────
             0
```

41

GA1504

Domino Dare I

Cut out the dominoes on the bottom of the page. Try to arrange them into the pattern shown here to form a correct multiplication problem. (A domino half with one dot stands for the number 1, two dots for the number 2, etc.)

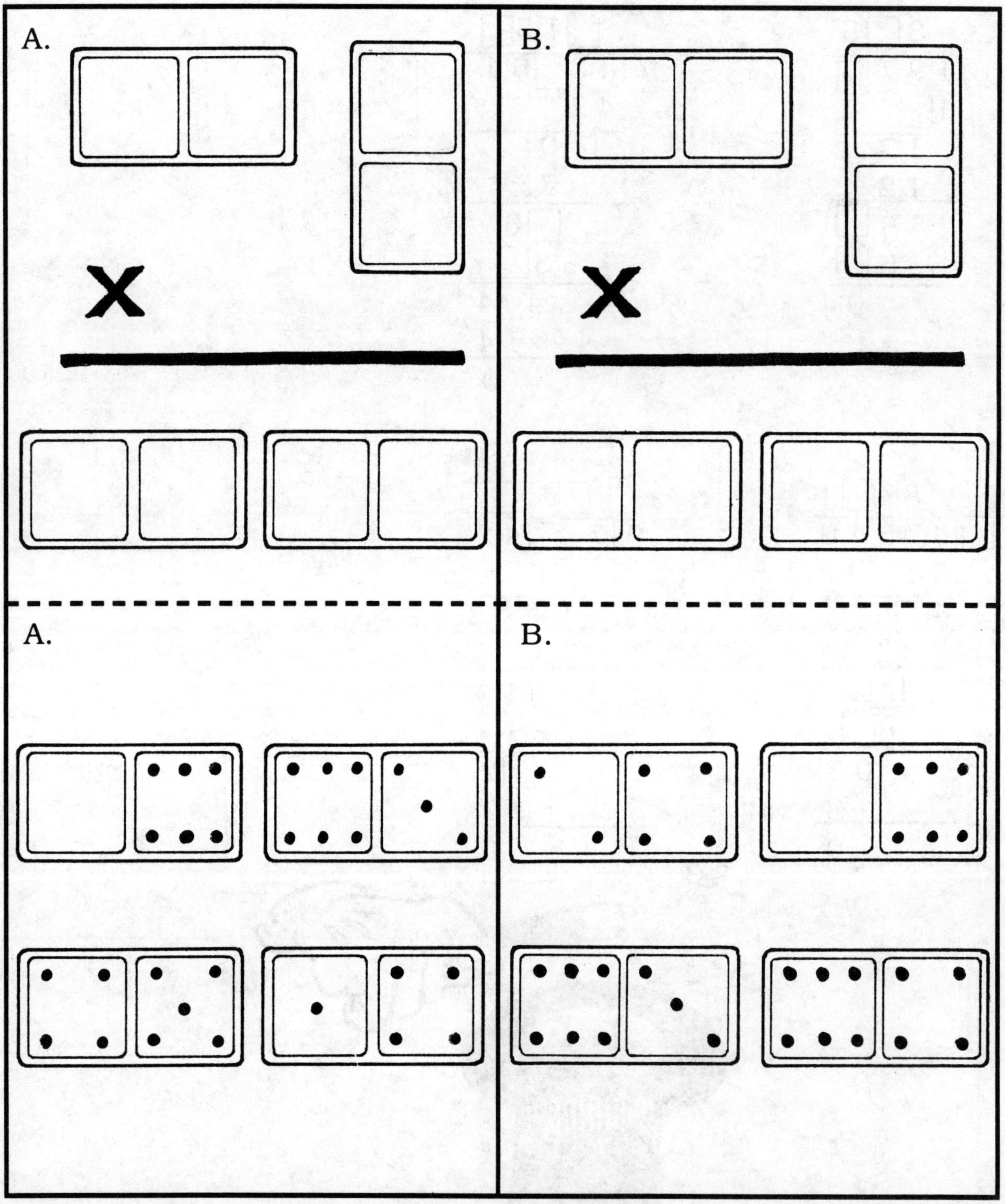

GA1504

Domino Dare II

This time, let's use dominoes from a double nine set. Again, cut out the dominoes on the bottom of the page. Try to arrange them into the pattern shown here to form a correct multiplication problem.

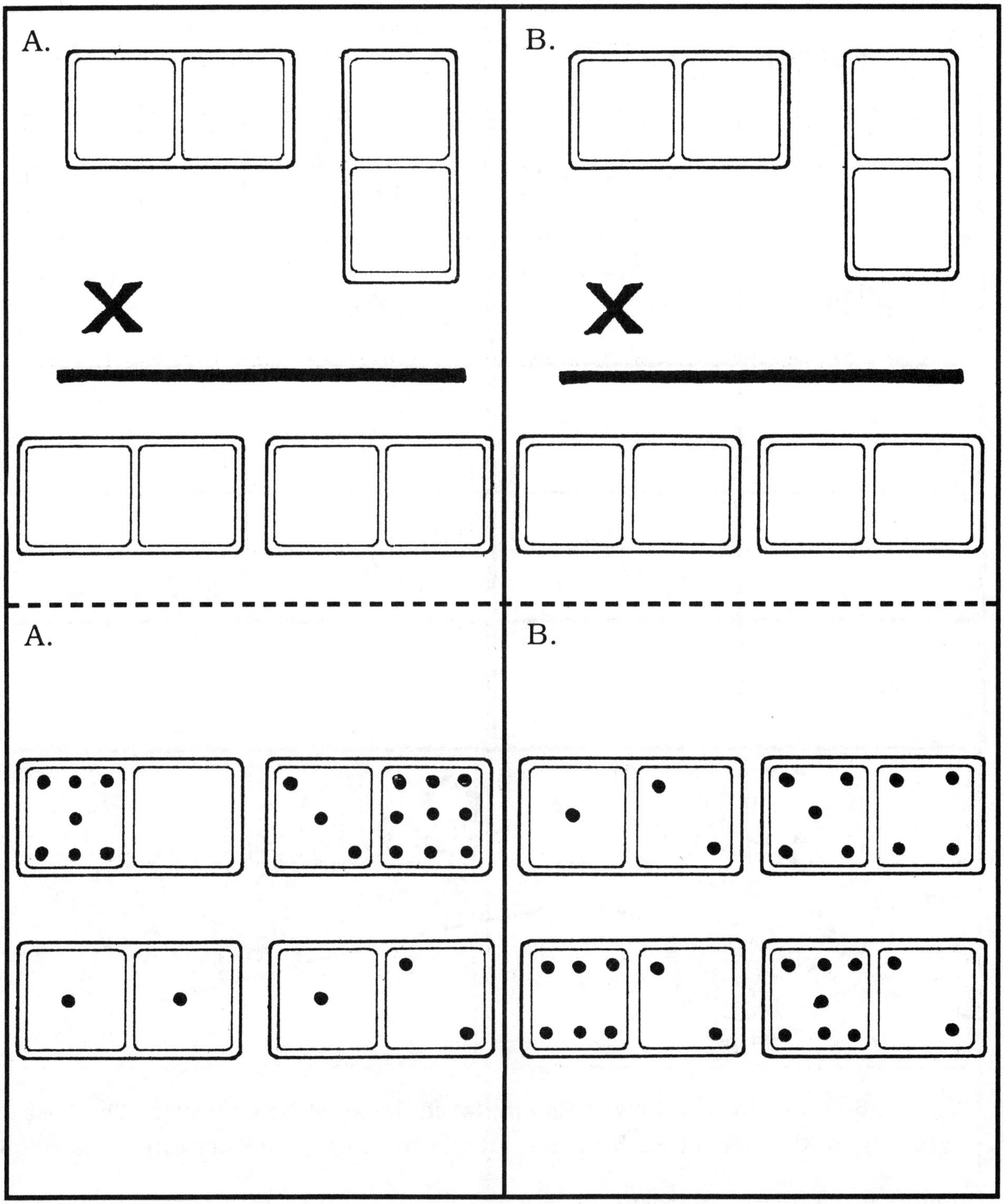

A.

B.

A.

B.

Rewarding Work

Mr. Boss wants to give a special reward to the employee who worked the most hours last year. First, study the chart below and circle the name of the person you think worked the most hours. Then calculate the actual hours each worked.

Who worked the most hours? _____

Name	Hours Per Week	Number of Weeks	Total Number of Hours
Bob	42	48	
Shari	46	42	
Mary	37	51	
Randy	38	50	

**The plaque Mr. Boss will give this employee needs to be engraved. The cost of engraving is 35¢ per letter. Mr. Boss needs to have 32 letters engraved. How

much will the engraving cost? _____

GA1504

Table Time

Finish this multiplication table by working back and forth between multiplication and division problems.

X	17		49	78
56	952	1,288		
90				
34		782		2652
	1054		3038	

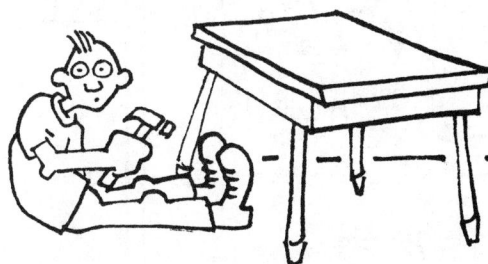

NOT **THAT** KIND OF **TABLE!**

**Make your own completed multiplication table. Then copy it for a friend, leaving blanks for him or her to fill. Be sure to provide enough information for your friend to be able to complete it.

(Teacher: This activity could be completed or checked with a calculator.)

GA1504

Room to Work

This is a picture of Betsy's dollhouse. It has four rooms, all of the same size, so each room is $1/4$ of the dollhouse. Cut out the furniture below and glue it in the right place.

Kitchen–bottom left $1/4$
Bedroom–top left $1/4$
Living room–bottom right $1/4$
Bathroom–top right $1/4$

Four Farmers

Here are two farms:

1. Farmer John owns $^1/_3$ of a farm that has cows on it. Put a *J* on John's farm.

2. Farmer Russ owns $^1/_2$ of a farm that has apple trees on it. Put an *R* on the farm that belongs to Russ.

3. Farmer Dale owns $^1/_3$ of a farm that has corn on it. Put an *D* on Dale's farm.

4. Farmer Phil owns $^1/_2$ of a farm that has corn. Put a *P* on Phil's farm.

GA1504

Just Clowning Around!

You have been asked to help eight clowns get ready for the circus. Draw features on the clown faces below so that:

a. $^3/_8$ of the clowns have red hair.

b. $^1/_8$ of the clowns have green hair.

c. $^4/_8$ of the clowns have orange hair.

d. $^5/_8$ of the clowns have hats.

e. $^7/_8$ of the clowns have large, round eyes.

f. $^1/_8$ of the clowns have small, triangular eyes.

g. $^2/_8$ of the clowns have frowns.

h. $^6/_8$ of the clowns have big smiles.

i. $^8/_8$ of the clowns have red noses.

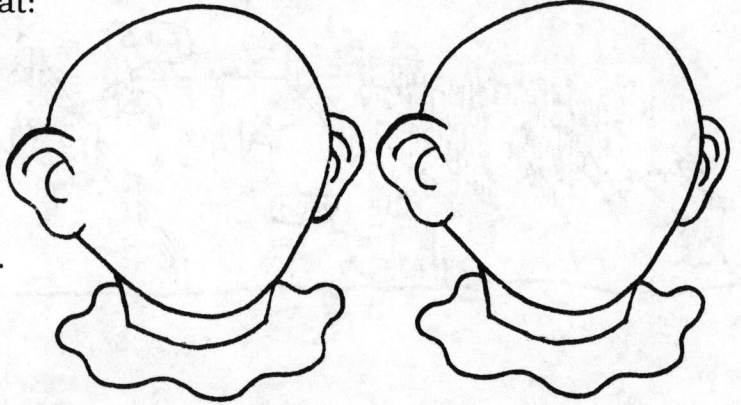

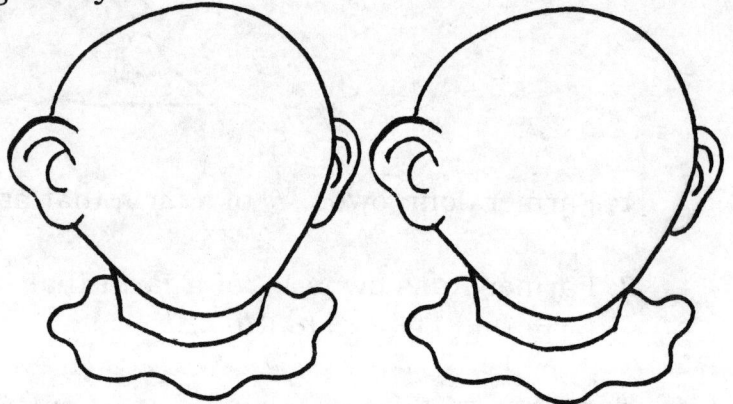

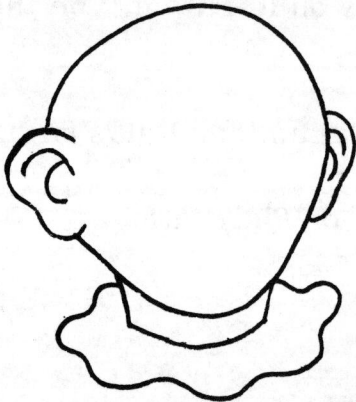

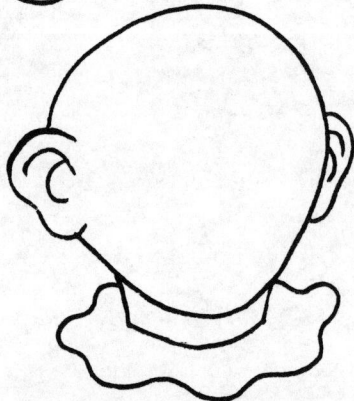

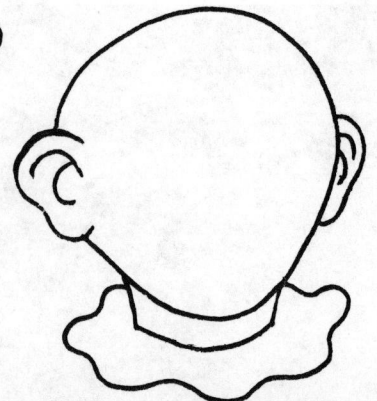

GA1504

Fractional Sports

The names of four sports can be found in the fractional parts of words below. Write the sport in the blank that can be spelled in each box.

1. Use the first $^2/_5$ of HORSE. Use the last $^2/_5$ of BLOCK. Use the last $^2/_5$ of MONEY. Sport: _____	2. Use the last $^4/_5$ of MARCH. Use the first $^2/_3$ of ERA. Use the first $^1/_6$ of YELLOW. Sport: _____
3. Use the first $^1/_2$ of BOOT. Use the last $^1/_2$ of TAXI. Use the last $^2/_5$ of SLING. Sport: _____	4. Use the last $^1/_2$ of MASK. Use the first $^1/_3$ of ILL. Use the first $^2/_3$ of INN. Use the first $^1/_4$ of GOAT. Sport: _____

Next, we'll give you the sport, and you fill in the fractional part of each word that should be used.

5. Sport: CROQUET Use the first _____ of CROAK. Use the first _____ of QUILT. Use the last _____ of SWEET.	6. Sport: TENNIS Use the last _____ of OFTEN. Use the first _____ of NEXT. Use the last _____ of THIS.

Now you're completely on your own. Try to write your own fractional words to spell one of these sports: soccer, baseball, rugby, swimming, wrestling, basketball.

 GA1504

Star Search

Can you help this space alien find his spaceship before it blasts off? The alien can only pass through fractions that are equal to $\frac{1}{2}$. Find a path from the alien to his ship that will lead you only through fractions that are equivalent to $\frac{1}{2}$.

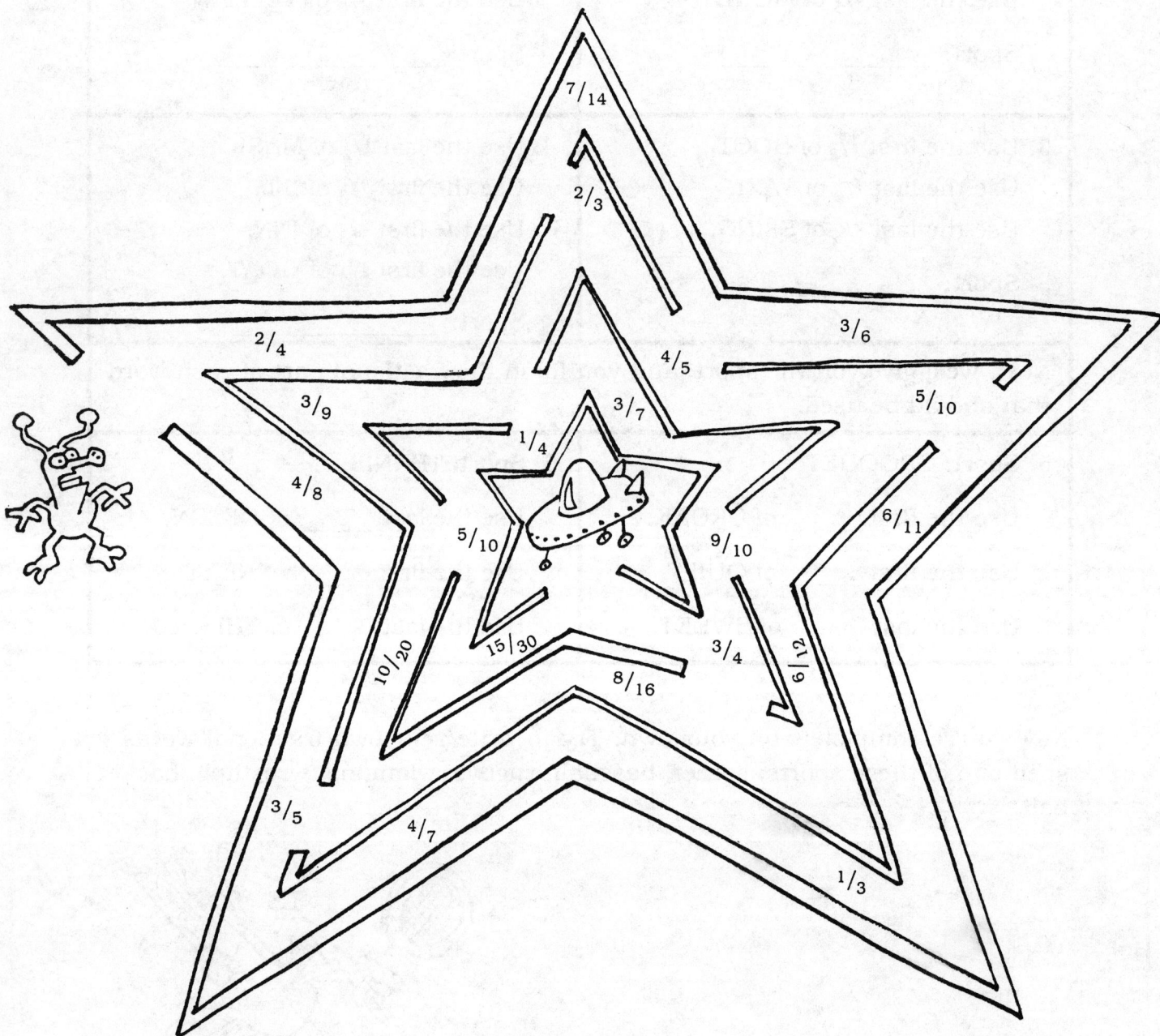

7/14
2/3
2/4
3/6
4/5
3/9
5/10
3/7
1/4
4/8
5/10
9/10
6/11
10/20
15/30
3/4
6/12
8/16
3/5
4/7
1/3

GA1504

Shady Fractions

Add the 2 fractions in each area below.
- If the sum is greater than 1, leave the space blank.
- If the sum is less than 1 or equal to 1, shade the area.

When you're done, you will discover a familiar figure.

$1/2 + 1/2$

$4/5 + 1/5$

$5/8 + 1/8$

$1/4 + 1/4$

$3/5 + 3/5$

$2/9 + 4/9$

$7/10 + 2/10$

$4/7 + 5/7$

$5/11 + 3/11$

$4/11 + 5/11$

$5/12 + 7/12$

$1/3 + 1/3$

$7/9 + 4/9$

$3/4 + 1/4$

$3/7 + 3/7$

$1/6 + 5/6$

$1/5 + 2/5$

$9/10 + 3/10$

$5/9 + 7/9$

$11/15 + 2/15$

$4/5 + 2/5$

$3/8 + 7/8$

$5/8 + 3/8$

$7/9 + 1/9$

$9/10 + 1/10$

Pocket Change

Ryan has 12 coins in his pocket.

- $1/2$ are dimes.
- $1/4$ are nickels.
- $1/6$ are quarters.
- $1/12$ are pennies.

How much money does Ryan have? _____

•• Suppose that Ryan had 24 coins in his pocket and the same fractions were

true. How much money would Ryan have? _____

•• Now suppose that Ryan had 36 coins and the same fractions were still true.

How much money would Ryan have? _____

CLINK!

CHA-CHINK!

Color Graph

Mrs. Calvert's first grade class voted to find out everyone's favorite color. Here is a list that shows how many students voted for each color:

Red–6 students
Blue–8 students
Green–2 students
Purple–10 students
Yellow–4 students

Color in the bars on this graph to show the same information. Use the same colored crayons as named in the list.

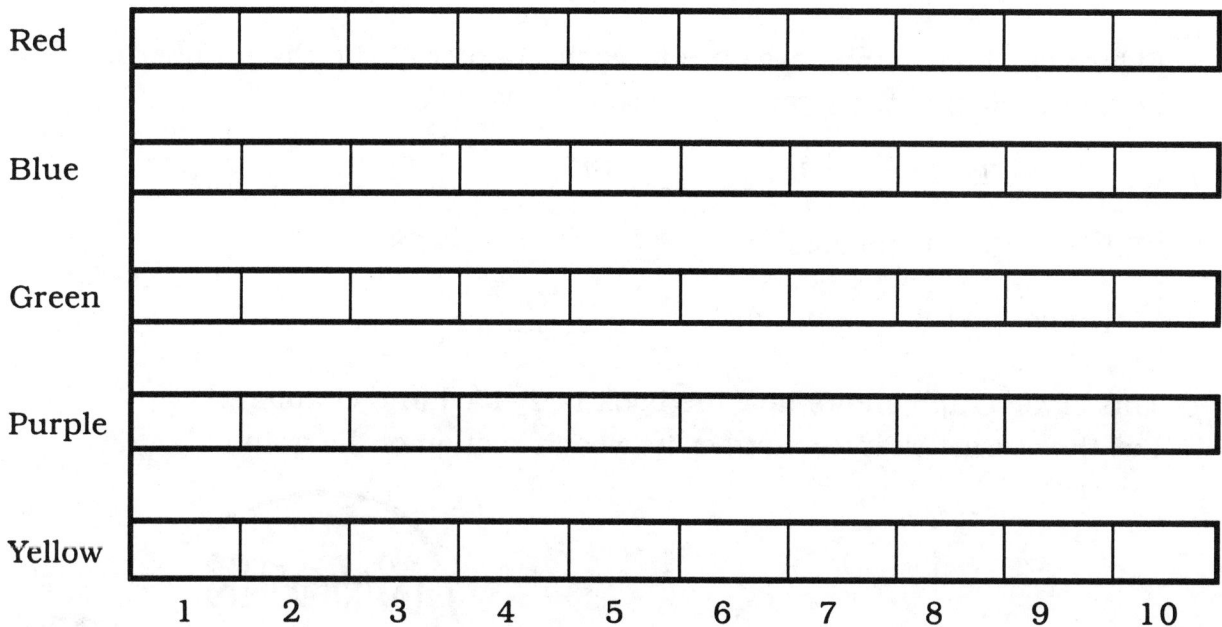

Red

Blue

Green

Purple

Yellow

1 2 3 4 5 6 7 8 9 10

GA1504

Laugh a Little!

Here is the table of contents for a joke book. It has 5 chapters. Each chapter begins on the page shown and ends on the page just before the one where the next chapter begins.

**Table of Contents
for
My Very Own Joke Book**

a. Figure out how many pages are in each chapter. Write the number of pages next to each chapter here:

I. _____   II. _____   III. _____   IV. _____   V. _____

b. Which chapter is the longest? _____

Which chapter is the shortest? _____

c. This circle graph shows how the book is divided into chapters.
Put the correct chapter number into each section of the graph.

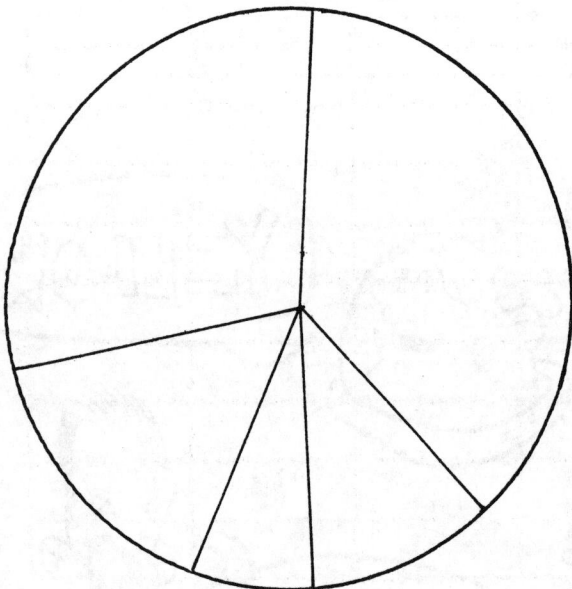

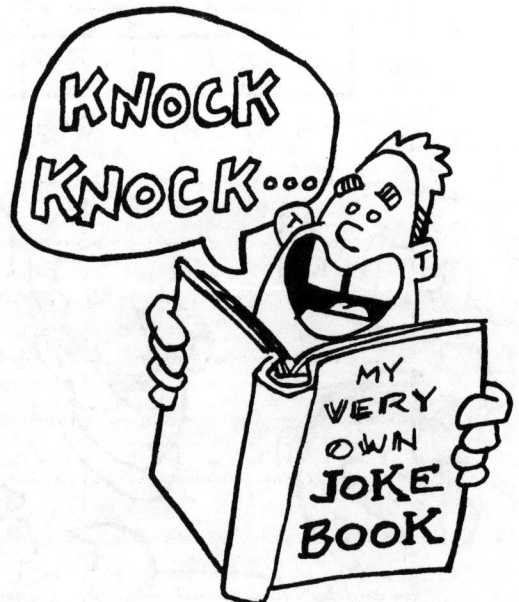

Correct the Collections

Mr. Lake's 4th grade class took a survey of what his students like to collect. They recorded the answers on this chart.

Collections	Number of Students Who Have One
Model dinosaurs	6
Model cars and trucks	4
Dolls and stuffed animals	9
Stamps	3
Rocks	5
Baseball cards	10
Posters	4

Someone then made a graph of the same information, but they made several mistakes. Circle the ones you find and write in the corrections.

What Our Third Graders Like to Collect

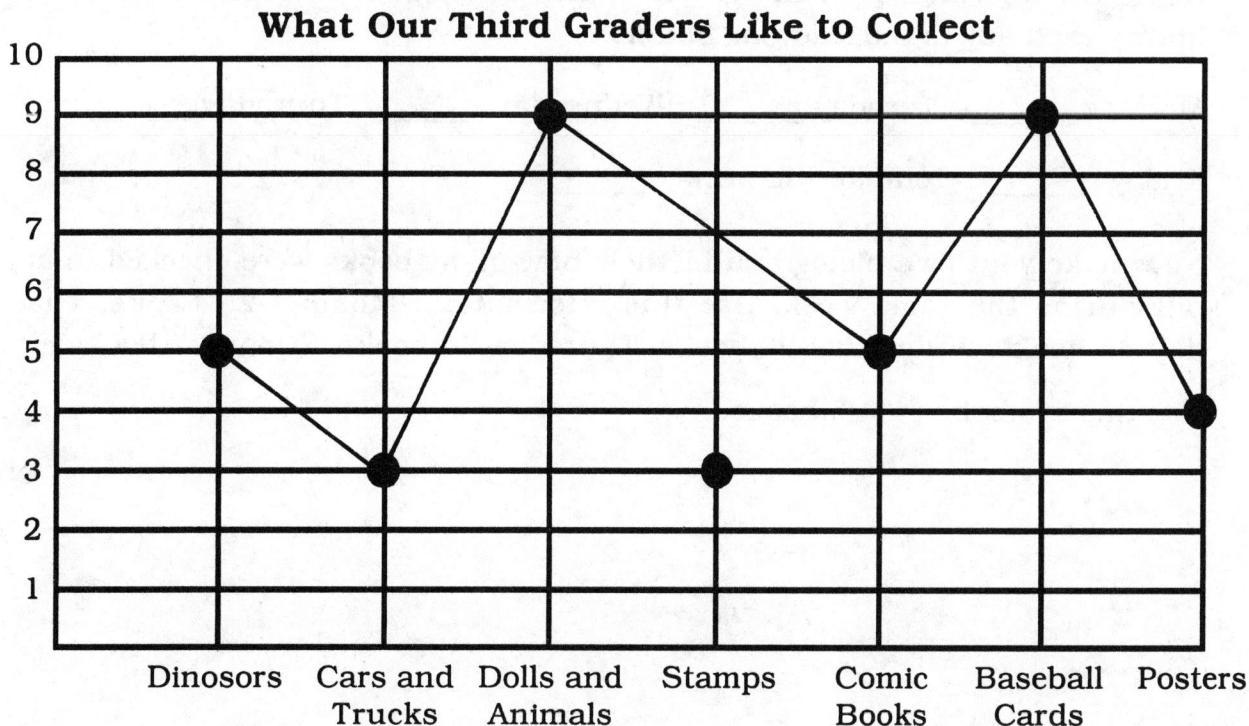

Reading and Making a Pictograph,
Counting by 5s and 10s,
Addition

Book Work

A lot of students use the library at Fisher Elementary School. This pictograph shows how many students used it in one week.

= 10 students

Monday

Tuesday

Wednesday

Thursday

Friday

Using the information in the graph, write the number of students using the library each day in the following blanks:

Monday _____   Tuesday _____   Wednesday _____   Thursday _____

Friday _____   Total for the week: _____

Now make your own pictograph to show how many books were checked in and out during the same week. Use this information: Monday–25 books, Tuesday–45 books, Wednesday–20 books, Thursday–50 books, Friday–30 books.

Use this symbol: = 5 books.

Graph a Giraffe!

You can draw a giraffe by making and connecting dots on the graph on the next page. For each letter and number in parentheses (   ), find the place where those two lines meet and put a dot there. Then find the next dot and connect it to the first dot you drew. Remember to always connect the dots as you go, and cross out each set of parentheses (   ) as you use it. When you've finished the dots, add your own face and other details to the giraffe.

(D, 1)

(D, 9)

(D, 23)

(A, 23)

(A, 24)

(C, 26)

(E, 26)

(F, 28)

(F, 20)

(F, 13)

(N, 13)

(P, 10)

(O, 10)

(O, 11)

(N, 13)

(N, 1)

(M, 1)

(M, 5)

(L, 9)

(F, 9)

(E, 5)

(E, 1)

(D, 1)

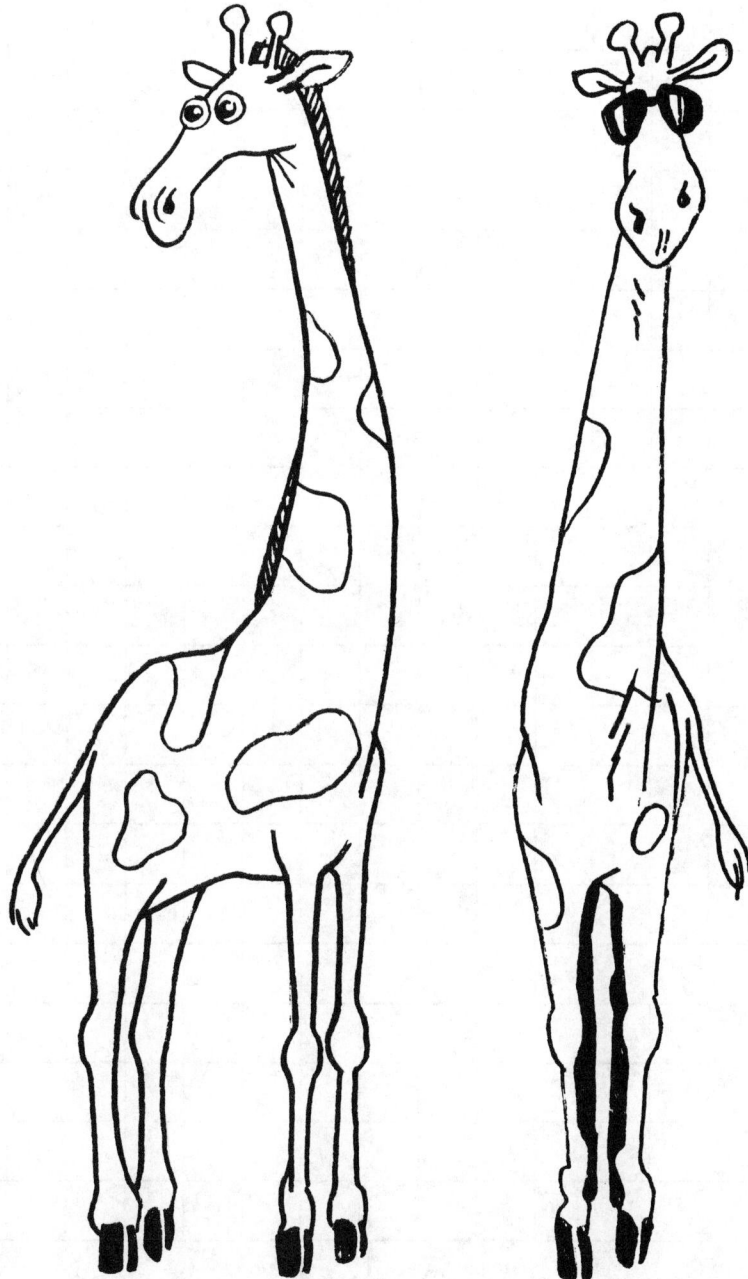

GA1504

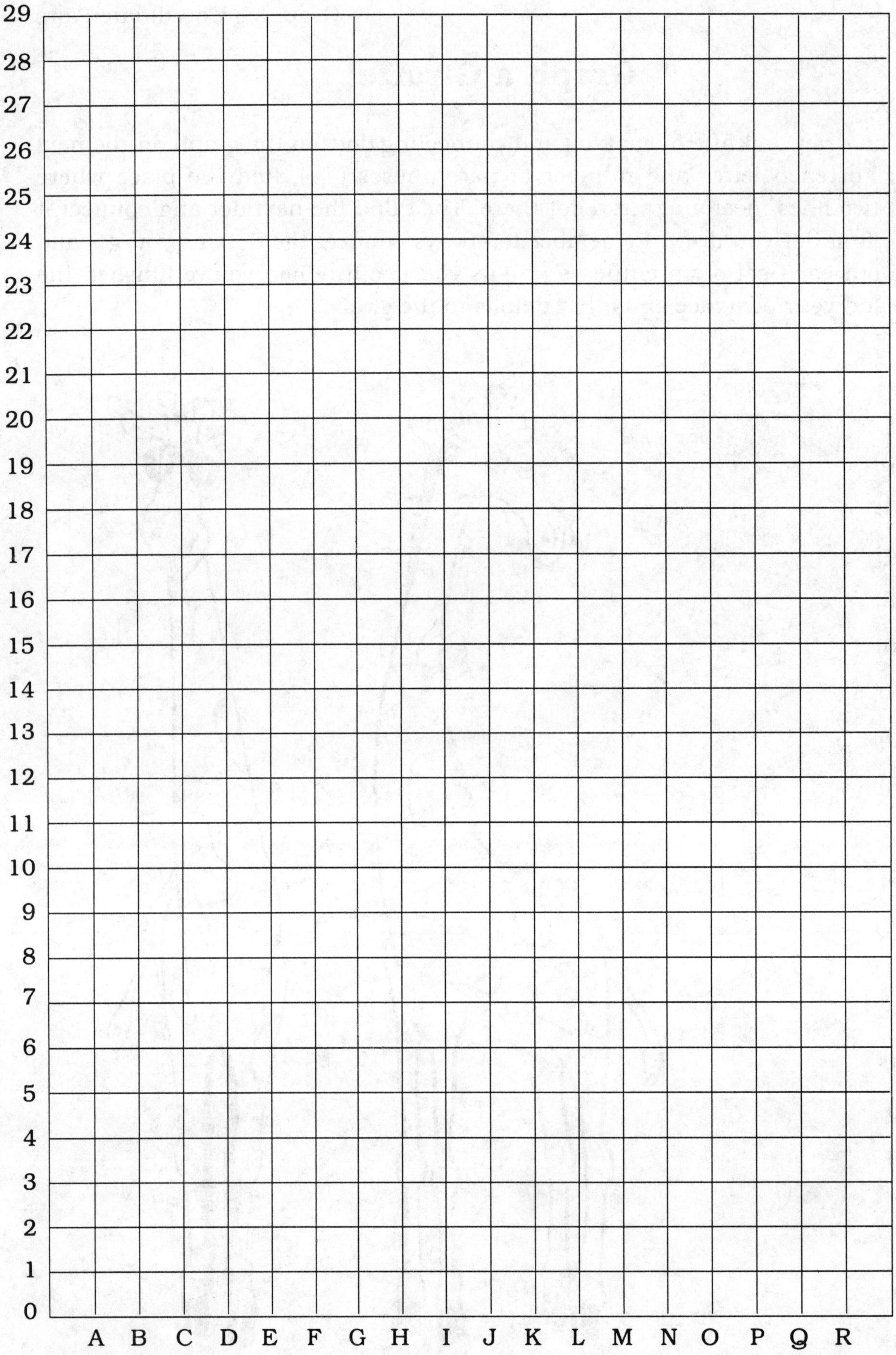
29 28 27 26 25 24 23 22 21 20 19 18 17 16 15 14 13 12 11 10 9 8 7 6 5 4 3 2 1 0
A B C D E F G H I J K L M N O P Q R

What Is It?

Draw a surprise picture by making and connecting dots on the graph on page 58. For each letter and number in parentheses ( ), find the place where those two lines meet and put a dot there. Then find the next dot and connect it to the first dot you drew. Remember to always connect the dots as you go, and cross out each set of parentheses ( ) as you use it. When you have finished the dots, you'll be headed in the right direction!

(K, 4)

(K, 1)

(G, 1)

(G, 4)

(C, 2)

(C, 6)

(G, 8)

(G, 10)

(C, 8)

(C, 11)

(G, 13)

(G, 22)

(B, 22)

(I, 29)

(P, 22)

(K, 22)

(K, 13)

(O, 11)

(O, 8)

(K, 10)

(K, 8)

(O, 6)

(O, 2)

(K, 4)

GA1504

What's Wrong?

Circle all of the mistakes on this calendar.

April

Sunday	Monday	Teusday	Thursday	Wedensday	friday	Saterday
	1	2	3	4	5	6
Ɫ	8	9	10	11	21	13
14	15	19	17	18	19	
S1	22	23	42	25	26	27
28	29	30	31			

GA1504

Becky's Birthday

Use these clues to find the month and day of Becky's birthday.

1. Becky's birthday month has less than 3 syllables.

2. Becky's birthday month has an *odd* number of days, but the day of her birthday is an even number.

3. Becky's birthday month does not end in *y* or begin with *A*.

4. Becky was born in the last 2 weeks of the month.

5. The date of Becky's birthday contains 2 digits which are the same.

Becky's birthday: _____ _____
                           (Month)   (Day)

Write the months here to help you.

GA1504

Calendar Computations

In what year will Halley's Comet probably next appear? Do these calculations to find the answers.

1. Find the number of days in September. _____

2. Add the number of days in July. _____ + _____ = _____

3. Multiply by the number of days in a week. _____ x _____ = _____

4. Multiply by the number of seasons in a year. _____ x _____ = _____

5. Add the number of days in a leap year. _____ + _____ = _____

6. Subtract the number of months in a year. _____ - _____ = _____

Answer: Halley's Comet should next appear in the year _____.

What's Next?

First write the name of the day, month, season, or number that comes next.
Then circle the letter in your answer shown by the number in parentheses ( ).
(For a 2, circle the second letter, etc.) Write the letters you circled in order in
the blanks at the bottom of the page. Read the message.

1. Sunday, Monday, Tuesday, _____ (9)

2. One, two, three, _____ (2)

3. Monday, Tuesday, Wednesday, _____ (3)

4. January, February, March, _____ (1)

5. Spring, summer, fall, _____ (6)

6. March, April, May, _____ (4)

7. June, July, August, _____ (9)

8. Five, six, seven, _____ (2)

9. Summer, fall, winter, _____ (6)

10. December, January, February, _____ (5)

11. Saturday, Sunday, Monday, _____ (1)

Message: __ __ __ __ __ __ __ __ __ __ __ !

Watch Match

Match each digital watch on the left to the clock on the right that shows the same time. Draw lines to connect your matches. You will have one digital watch left over.

1.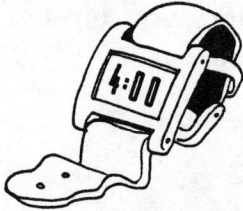

A.

2.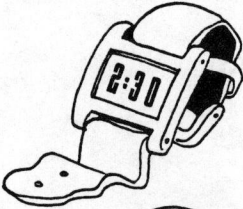

B.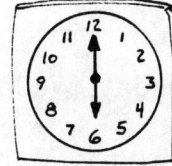

3.

C.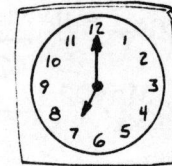

4.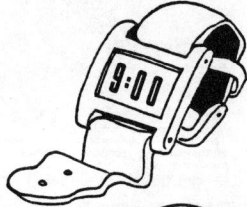

D.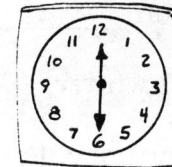

5.

E.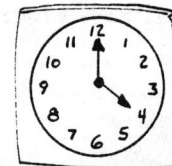

6.

F.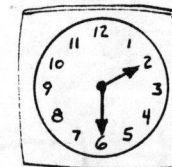

7.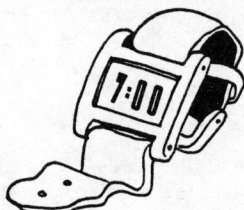

GA1504

Taxi Time

Here are all the things Tony the taxi driver has to do today. Can you help him straighten things out? Show him what to do first, second, third, and so on by numbering these activities from 1 to 12.

☐ 1:10 p.m.   Take Mrs. Pohutski to the library and wait for her.

☐ 10:35 a.m.   Pick up Mrs. Miller and take her to the grocery store.

☐ 4:15 p.m.   Take Mr. Rockhill and his cat to the vet.

☐ 9:15 a.m.   Pick up Mrs. Richmond at the airport.

☐ 11:30 a.m.   Get Mrs. Miller at the grocery store and take her home.

☐ 5:00 p.m.   Leave work.

☐ 2:05 p.m.   Drive Mrs. Williams to the post office, drug store, and dentist.

☐ 9:00 a.m.   Report to work.

☐ 10:55 a.m.   Pick up Mr. Fisher at the doctor's office and then take him home.

☐ 3:30 p.m.   Pick up Dr. Jacobs at the airport.

☐ 12:00 Noon Lunch break

☐ 9:45 a.m.   Pick up Mr. Fisher at his home and take him to the doctor.

Clock Calamity

Help! One dozen numbers just fell off this Roman clock, and now one of the numbers is missing. First figure out the missing numeral and write it here.

_____

Next, write all 12 Roman numerals in their correct places on the clock.

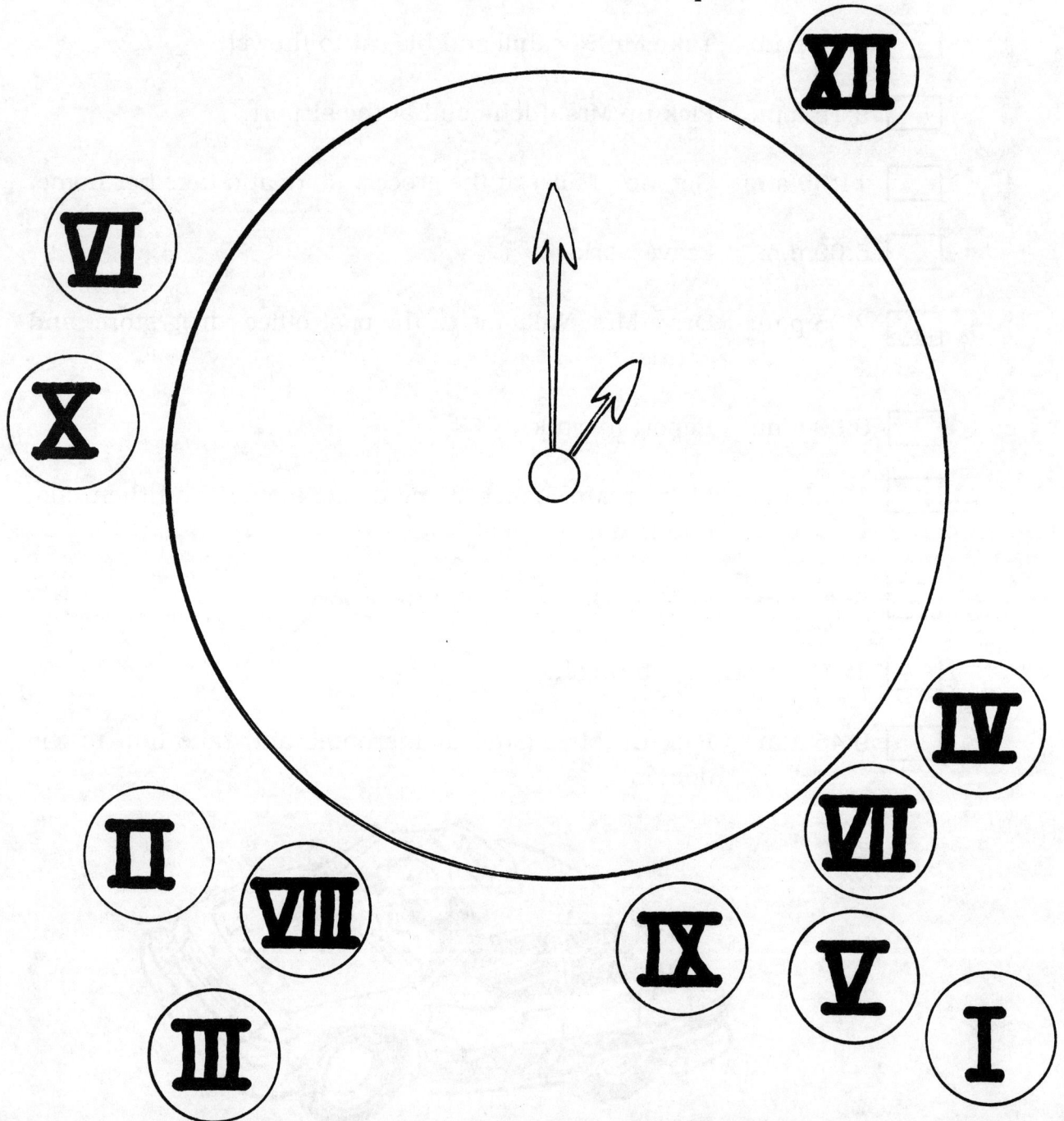

More or Less

Circle the Roman numeral in each pair that's worth more.

1. III   IV

2. VI   IV

3. XI   XV

4. XL   XXXV

5. XI   IX

6. XX   XV

7. LV   LX

8. VI   IX

9. XIX   XVI

10. LII   LXX

Now compare all the Roman numerals listed above. Which one is worth the

most of all? _____

Shape Up!

Which picture could you make by using every shape in this set? (Don't use any extras.) Draw a line under your answer.

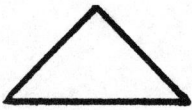

A.

B.

C.

D.

68

Shape Art

Make your own interesting picture using 3 circles, 3 triangles, 3 rectangles, and 3 squares.

Count the shapes in your picture to make sure you have three of each shape. Color or decorate your picture.

(Teacher: Students could also complete this page by cutting out shapes and pasting them in the box.)

GA1504

Tricky Triangles

How many of the triangles at the bottom of this page do you think will fit inside

the box? _____

Now cut out the triangles. See how many will fit completely inside the box. You

may turn or flip the triangles. How many went inside? _____

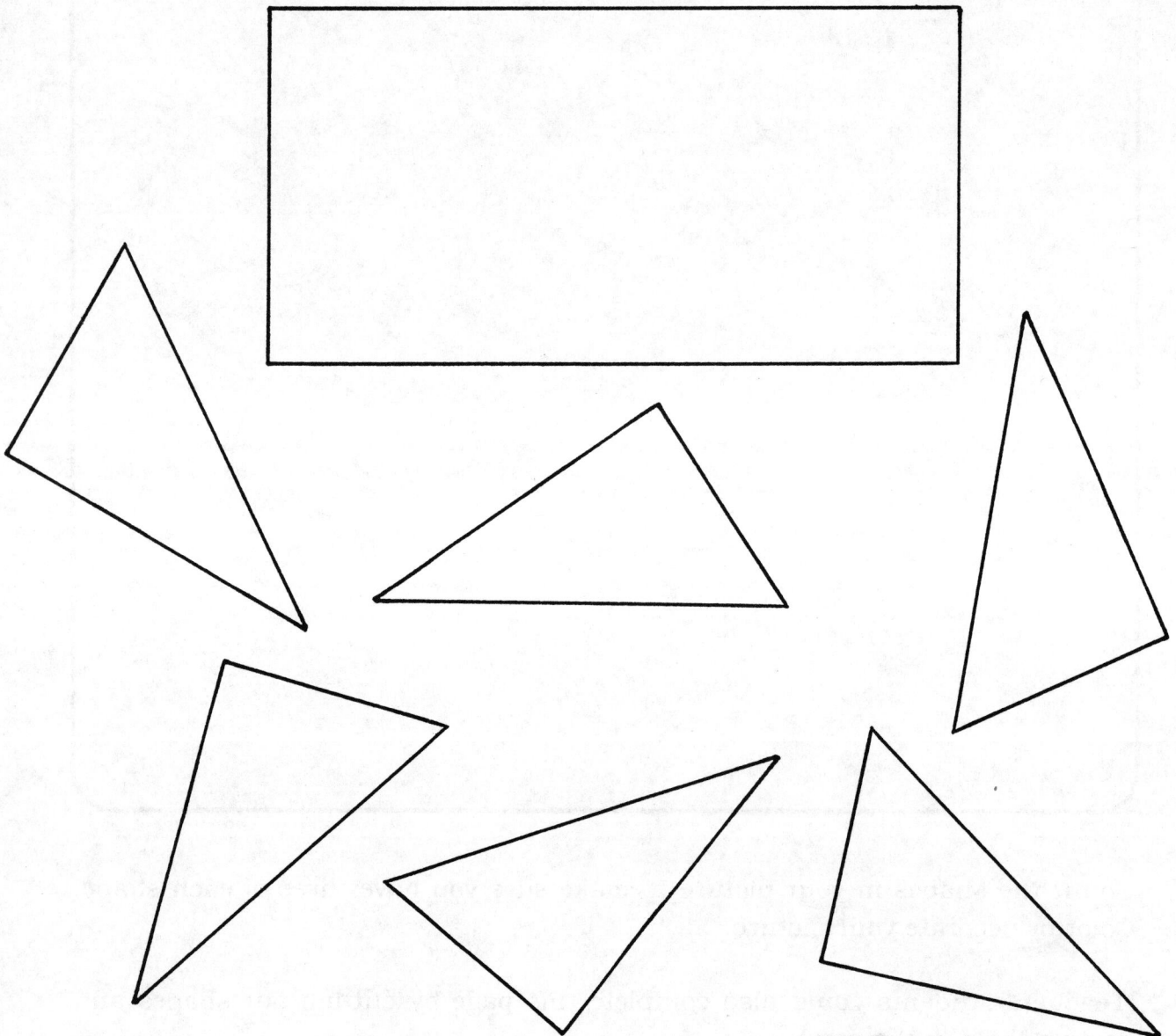

Common Couples

Each shape here has something in common with one of the others. Draw lines to connect each pair.

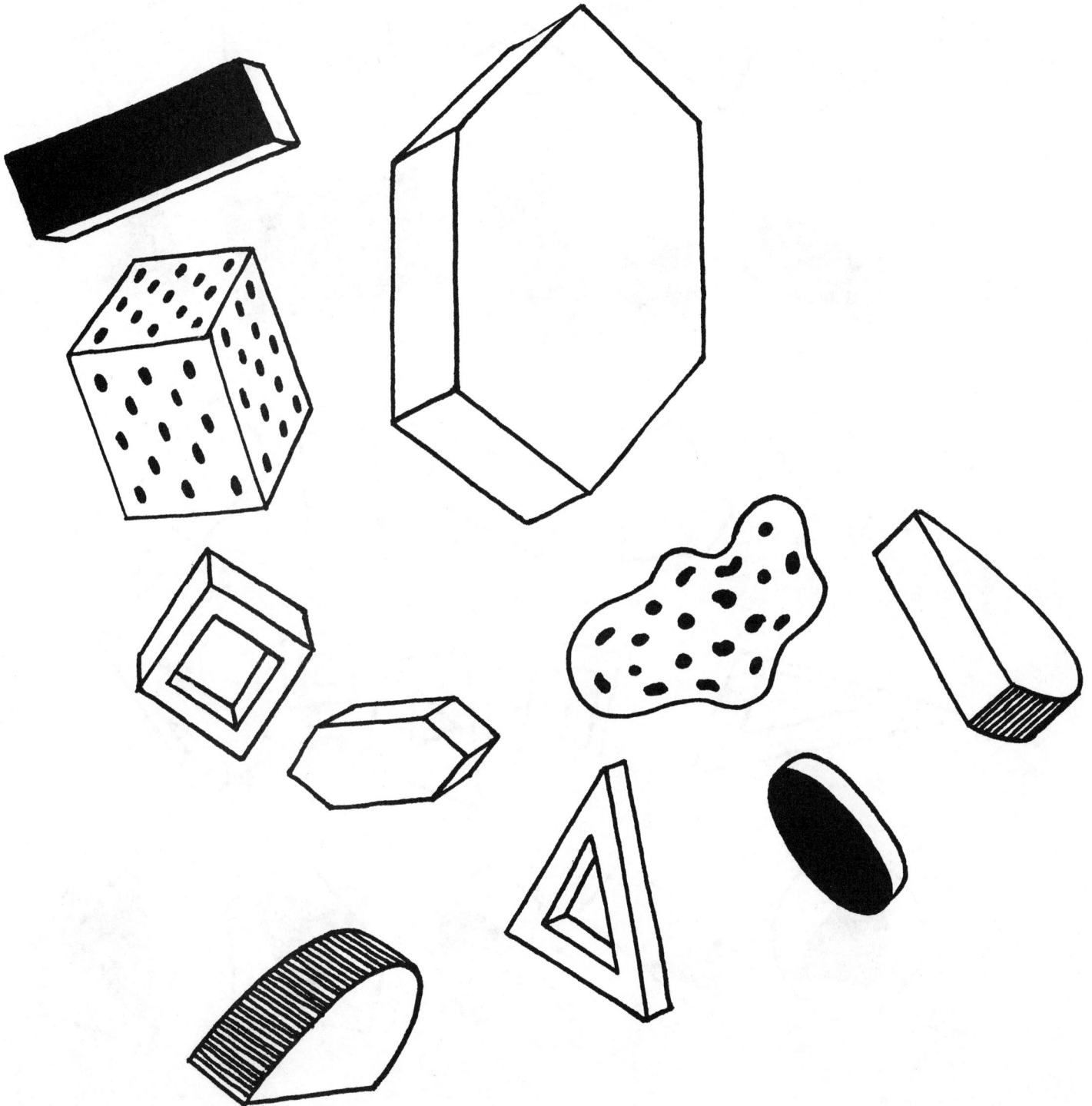

GA1504

Shape Mistake

In each row, one shape doesn't belong. When you figure out which one it is, cross it out.

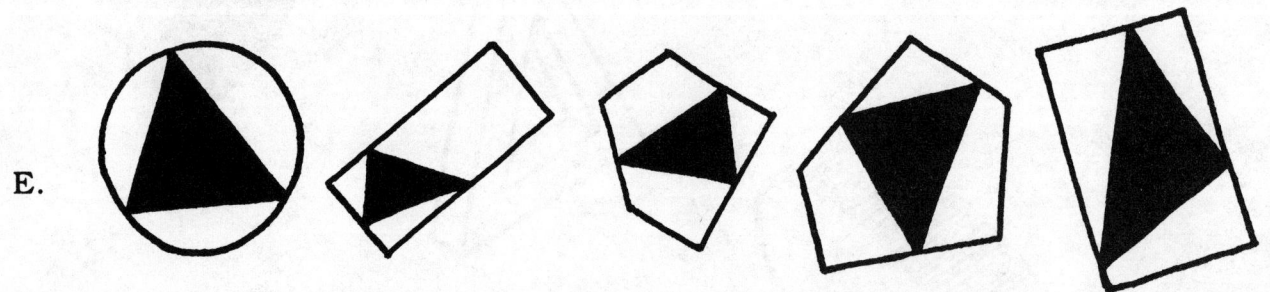

A.

B.

C.

D.

E.

Folding Fits

Matt and Amy are packing away their winter clothes. They noticed that some items, when folded in the middle, make two halves that fit on top of each other. For example, when they folded this shirt on the dotted line,

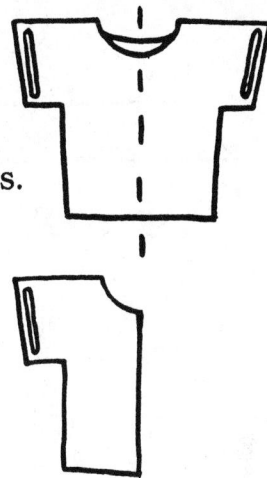

they made two halves which, stacked together, looked like this.

They also noticed that some things like this mitten can't be folded into two matching halves.

Figure out which items below can be folded into halves (like the shirt). Draw in a dotted line to show where they can be folded. Put an X on the things that can't be folded into halves.

**One item can be folded two different ways. Circle it and draw in both folding lines.

GA1504

Baffling Blocks

Baby Elin is putting these 3 blocks into this shape box. If she always has the arrow in front, how many different ways can Elin put each block in its spot? (Think about the directions in which the arrow can point.) Write your answers below.

A

B

C

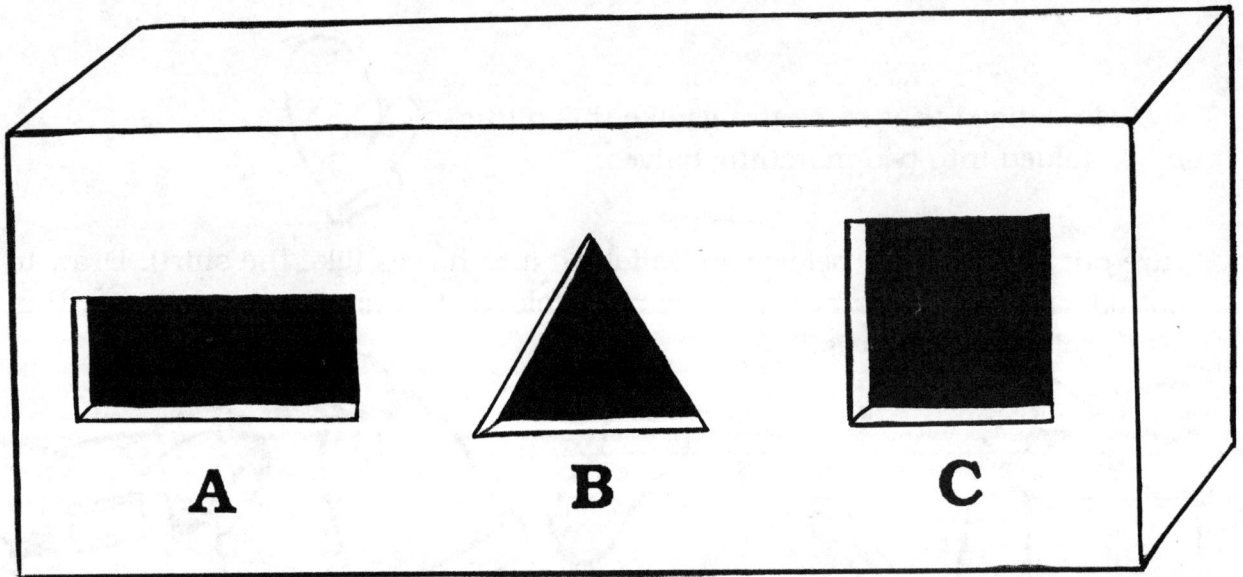

Block A can go in the box _____ different ways.

Block B can go in the box _____ different ways.

Block C can go in the box _____ different ways.

GA1504

Square Puzzlers

Cut out each shape, one at a time, and separate it into pieces by cutting on the dotted lines. Rearrange the pieces into the square.

A.

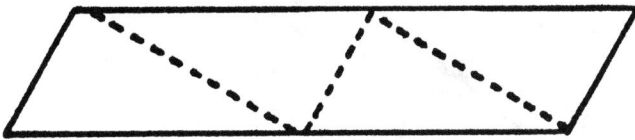

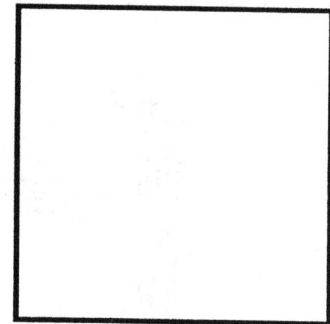

Sketch your solutions in the squares.

B.

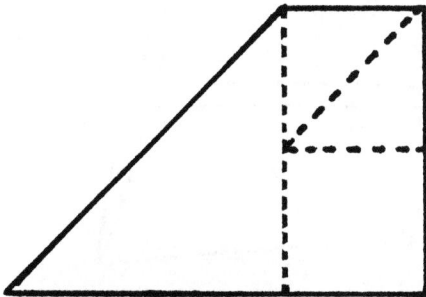

C.

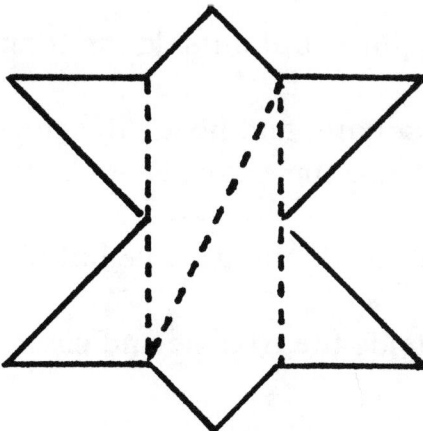

Now it's your turn to design. Cut this last square into 4-6 pieces. Rearrange them into another shape. Glue your new shape here, or make a drawing of it.

GA1504

A Neat Drawer

Keith is a very tidy person. Can you help him put some things away? Cut out the objects below. Follow the directions to find where each one belongs. Glue or sketch each item in its correct place.

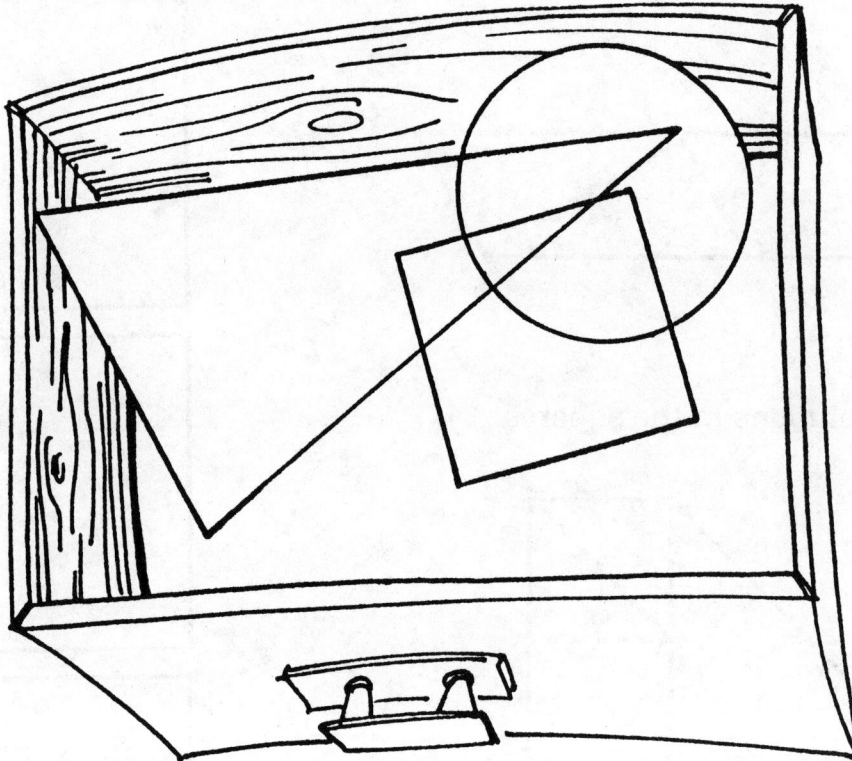

1. Put Keith's socks inside the drawer inside the triangle but outside the circle and outside the square.
2. Put the belt inside the circle, inside the square, but outside the triangle.
3. Put the slippers outside the drawer.
4. Put the tie inside the triangle, inside the square, and inside the circle.
5. Put Keith's wallet inside the circle, outside the triangle, and outside the square.
6. Put the sunglasses inside the square and inside the triangle but outside the circle.
7. Put the T-shirt inside the square and outside the triangle and circle.

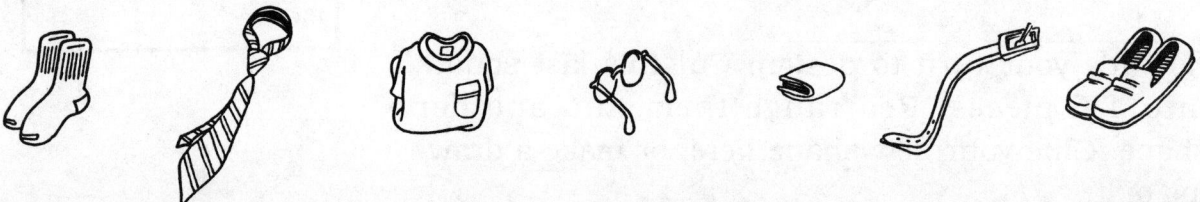

GA1504

Wall Painting

Mrs. Decor has just had this wild design drawn on her living room wall. Now she's ready to paint it. She wants to be sure the shapes that touch one another are painted with different colors, yet she wants to use the fewest number of colors possible. Try this with your crayons or colored pencils.

How many colors will Mrs. Decor need? _____

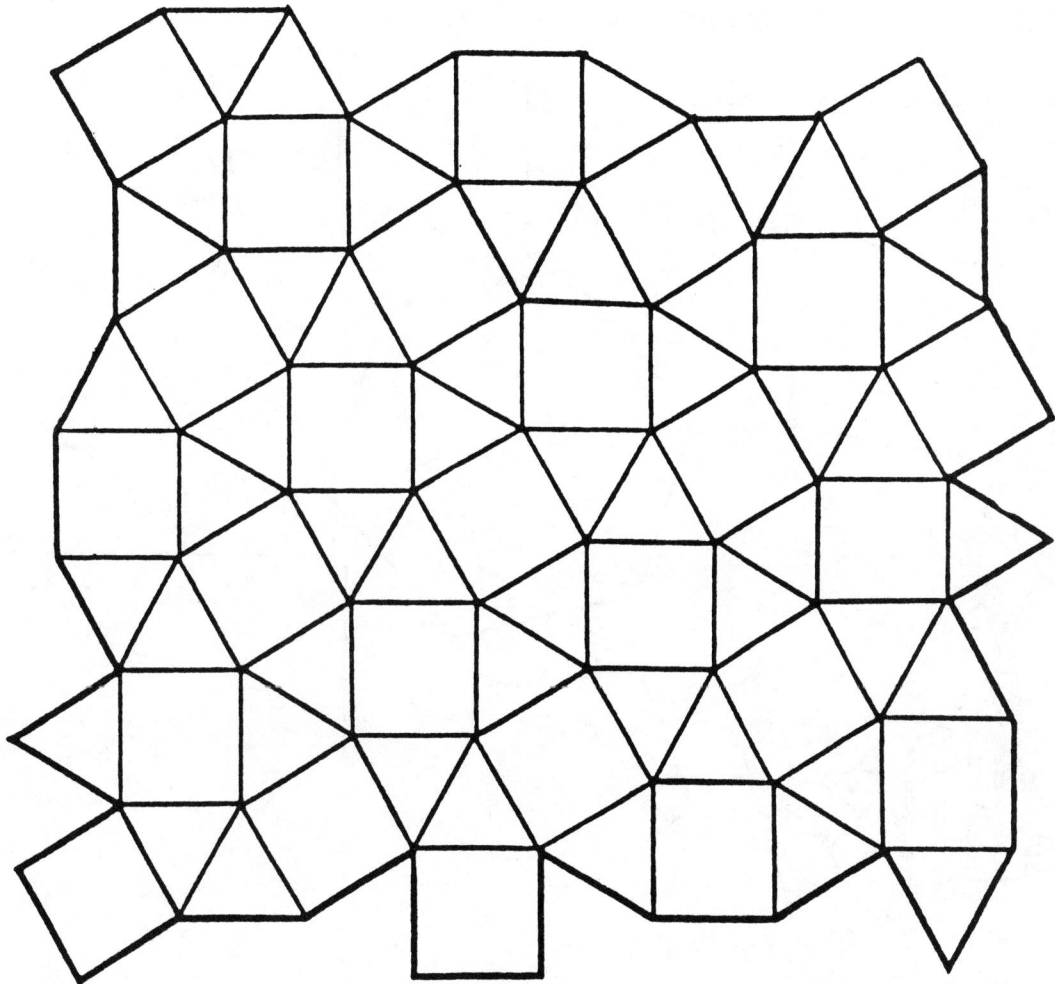

GA1504

Floor Samples

Mrs. Decor is recovering some of the floors in her house. She has designed several floor tiles shown here. She's trying to decide which ones will work best. Some tiles will fit together perfectly without overlapping or leaving spaces. Some tiles will not.

Here's how you can find out which ones fit properly:

1. Trace 3 or 4 copies of each shape.
2. Cut out your copies.
3. Try to fit them together inside the box.

Circle the shapes that work.

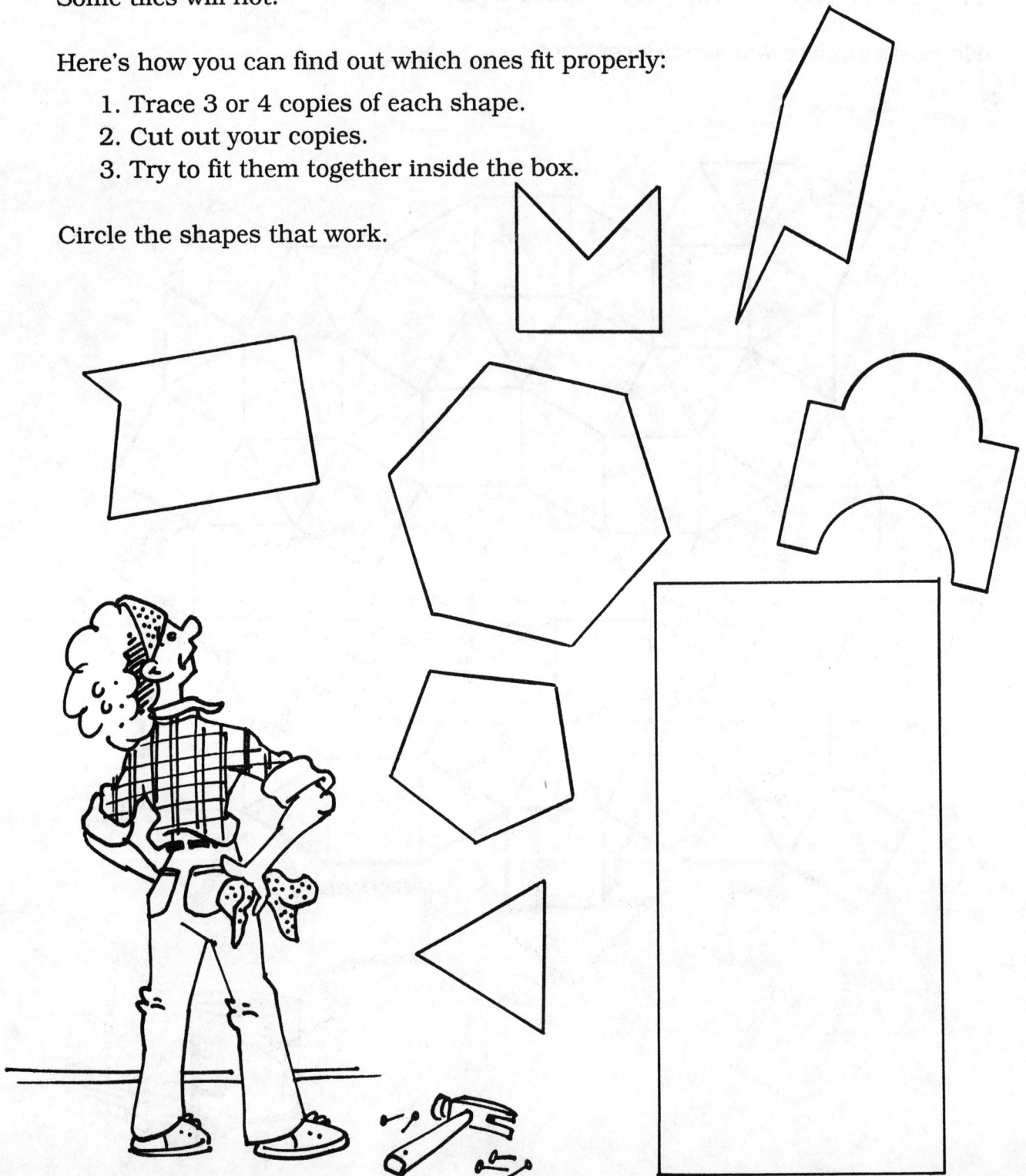

Shape Caper

First, study shape #1 carefully and circle the shapes below that you think are the same. Then trace and cut out shape #1. Place it on top of the others. Were your guesses right?

Which shapes are exact matches? _____

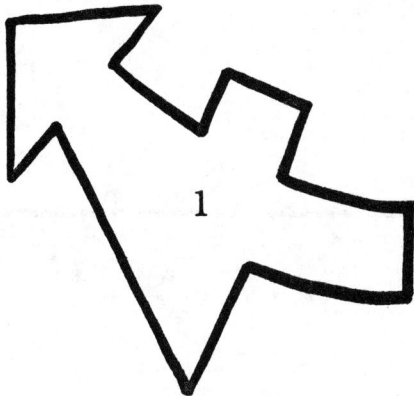

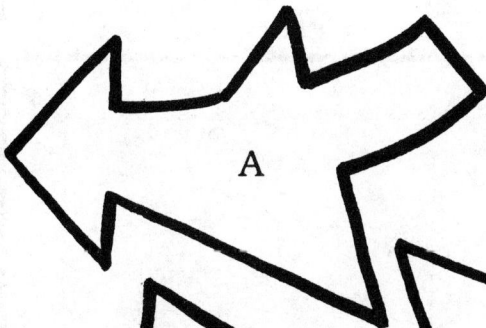

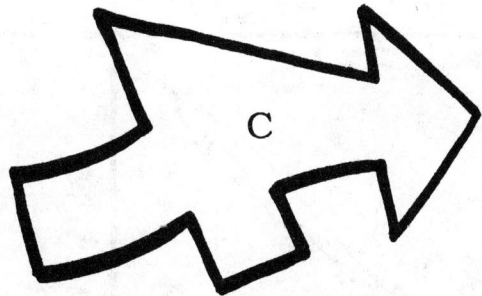

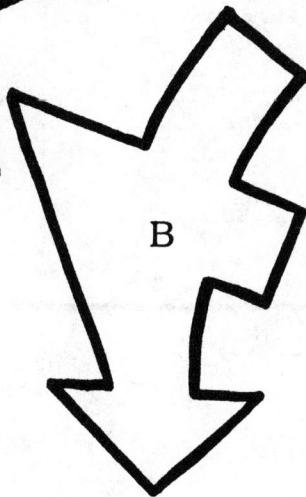

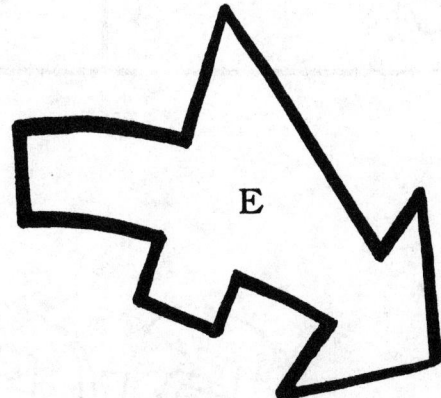

GA1504

Coin Cutouts

Cut out the coins at the bottom of the page. Place them next to each item below to show which coins you could use to buy the item. When you've found the answers, glue the coins in place.

comb 15¢	
apple 22¢	
pencil 8¢	

GA1504

Penny's Prices

In Penny's store, the price tags show the coins needed to buy each item. Look at these things for sale. Circle the one you think will cost the most. Then figure out the total price for each item and write it in the blank underneath. Was your guess correct?

_____ _____ _____

_____ _____ _____

GA1504

Rolling Coins

Two brothers, Nathan and Justin, just emptied their piggy banks. Here are the coins that fell out:

Nathan

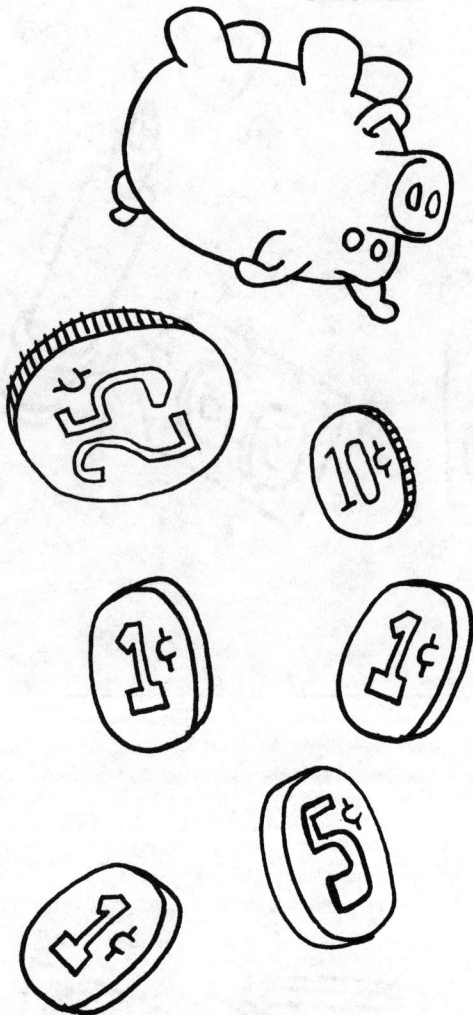

Justin

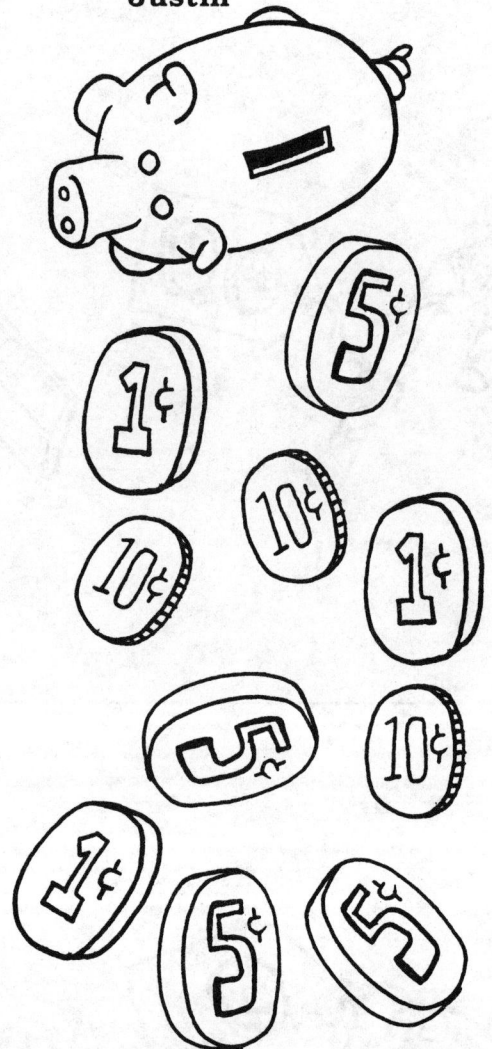

a. Earlier, the boys had a matching amount of money, but now they see that one coin must have rolled into the wrong pile. First, count how much money each boy has now and write it on the blank.

b. Then figure out how to make the piles worth the same. Finish this sentence:

To make both piles worth the same, you could move a _____ from

_____'s pile to _____'s pile.

GA1504

Barney's Banking

Barney the banker wants each customer to be happy with his service. When he gives out change, he is sure to give the customer exactly the number of coins requested. For example, if Mrs. Smith asks for 25¢ in 3 coins, Barney will give her 2 dimes and 1 nickel. How would Barney fill each request below? List the coins for each one on the lines provided.

1. 15¢ in 3 coins _____

2. 15¢ in 6 coins _____

3. 20¢ in 4 coins _____

4. 25¢ in 7 coins _____

5. 26¢ in 2 coins _____

6. 31¢ in 3 coins _____

7. 42¢ in 5 coins _____

8. 55¢ in 3 coins _____

9. 70¢ in 8 coins _____

10. 95¢ in 5 coins _____

GA1504

Barney's Bills

Barney the banker always gives his customers exactly what they want. For instance, if Mr. Smith asks to cash a $50 check and wants to receive exactly 8 bills, Barney hands him 2 $10 bills and 6 $5 bills.

How would Barney fill each request below? List the bills on the lines below.

1. $90 in 7 bills   _____

2. $30 in 5 bills   _____

3. $45 in 13 bills   _____

4. $85 in 8 bills   _____

5. $60 in 14 bills   _____

6. $70 in 7 bills   _____

7. $75 in 18 bills   _____

8. $100 in 7 bills   _____

9. $100 in 9 bills   _____

10. $100 in 16 bills _____

**What is the smallest number of bills needed to make $50? _____

**Without using one- or two-dollar bills, what is the largest number of bills you could use to make $50? _____

GA1504

Tools of the Trade

Carpenter Short's tools have become mixed up with his partner's tools. Help him find out which of these are his own. Carpenter Short knows all of his tools are more than 2 inches long and less than 4 inches long. Circle all the tools that belong to Carpenter Short.

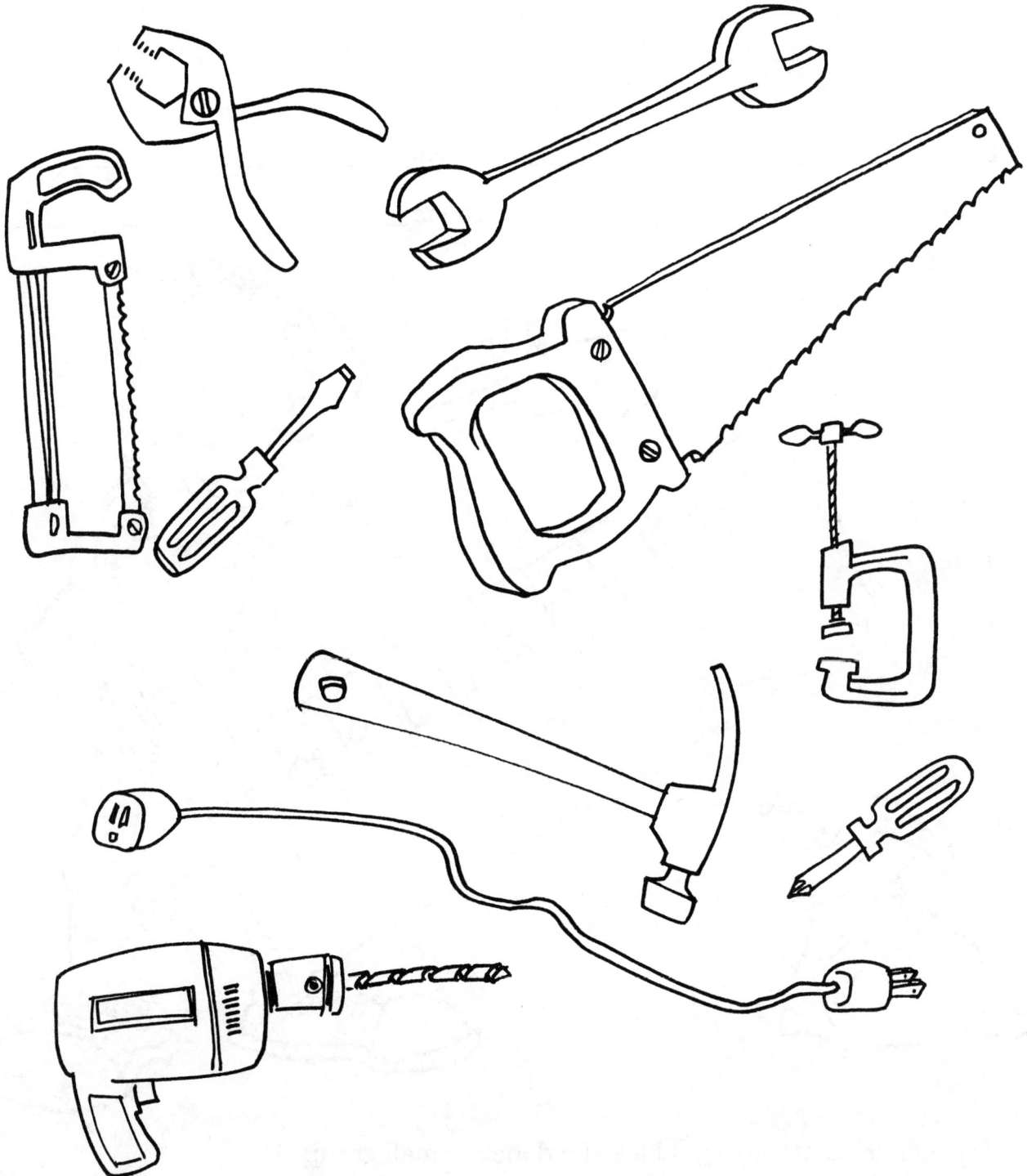

Scavenger Hunt

Look around your desk and room. Try to find something nearby that measures each length listed here. Draw a picture or write the name of each object you find on the blank next to its measurement. Use the ruler at the bottom to help you.

1 cm _____ 6 cm_____

2 cm _____ 7 cm_____

3 cm _____ 8 cm_____

4 cm _____ 9 cm_____

5 cm _____ 10 cm_____

(Teacher: This activity could also be done in small groups.)

GA1504

Mabel's Labels

The labeling machine at Mabel's Supermarket broke down. It stamped these groceries with the correct number portion of the weight, but without the unit of measurement. Finish labeling these items by writing either ounces (oz.) or pounds (lbs.) on each label to show their correct weights.

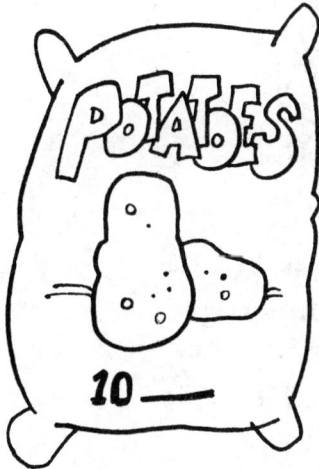

APPLES 3 ____

Granola Bars 12 ____

PEPPER 8 ____

POTATOES 10 ____

APPLE JUICE 64 ____

SALT 1 ____

HAM 6 ____

CEREAL 15 ____

PEA SOUP 11 ____

Laundry SOAP 5 ____

GA1504

Temperature Travels

*First, complete the thermometers to show the temperature given under each one.

*Next, locate these 4 cities on a U.S. map: Chicago, IL; New York City, NY; Miami, FL; and Tucson, AZ.

*Guess which thermometers show the average high temperatures in winter and in summer for each of the 4 cities listed. Write your guesses in the blanks. Check your answers with your teacher.

A. Winter   Summer

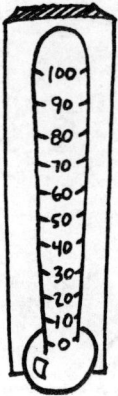 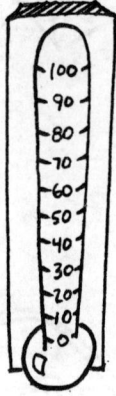

41° F   80° F

City: _____

B. Winter   Summer

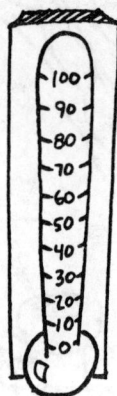 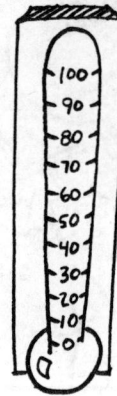

65° F   97° F

City: _____

A. Winter   Summer

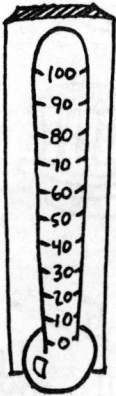 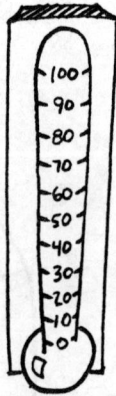

76° F   88° F

City: _____

B. Winter   Summer

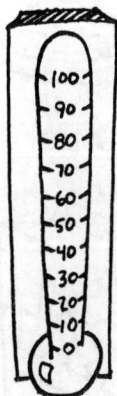 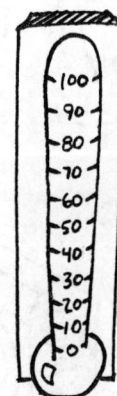

34° F   82° F

City: _____

Measurement Mess

These measurement labels belong on the objects below. Your job is to figure out which one goes where. Put the letter from each tag next to the object where it should be placed. When you're done, the items in the first column should all be measured by weight, those in the second column by length, and those in the third column by volume.

a. 5 cups	b. 30 gallons	c. $^1/_2$ cup
d. 3 feet	e. 2 tons	f. 3 ounces
g. 160 pounds	h. 2 quarts	i. 1 inch
j. 2 yards	k. 1 mile	l. 2 pounds

Weight	Length	Volume

GA1504

Tile Style

Mr. Parquet is recovering the floors in his house with square tiles. Imagine that this is the size of each floor tile: ☐ This is one square centimeter which is written 1 cm². Now imagine that each shape below is the floor of one of the rooms in Mr. Parquet's house. Find out how many tiles, or square centimeters, will fit in each room. Use a ruler to mark off each square unit. The first one is done for you.

1. <u>9 cm²</u>

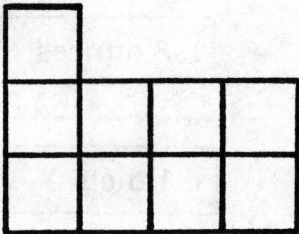

2. _____

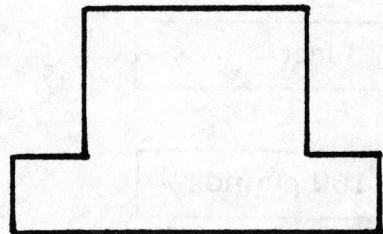

3. _____

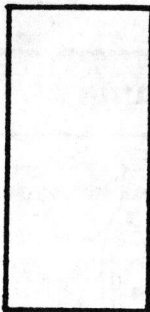

4. _____

5. _____

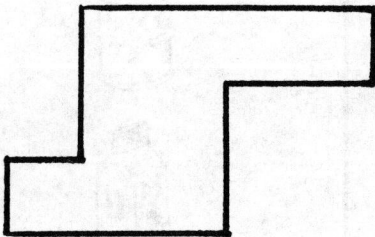

6. _____

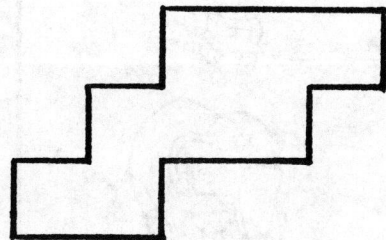

7. Draw your own shape with 10 cm².   8. Draw your own shape with 12 cm².

GA1504

Abbreviated Word Search

How well do you know measurement abbreviations? First write the complete unit of measure on each blank next to its abbreviation. (Watch your spelling!) Then find each word you wrote inside the word search. Words may appear up and down, across, or diagonally.

1. in. _____
2. ft. _____
3. yd. _____
4. mi. _____
5. m _____
6. km _____
7. cm _____
8. oz. _____
9. lb. _____
10. g _____
11. kg _____

12. c. _____
13. tsp. _____
14. tbsp. _____
15. pt. _____
16. qt. _____
17. gal. _____
18. ml _____
19. l _____
20. sec. _____
21. min. _____
22. hr. _____

```
N  O  O  P  S  E  L  B  A  T  R  M
Q  S  I  R  U  O  H  Y  T  E  E  N
E  E  U  N  A  R  A  M  T  T  T  R
T  C  E  O  C  R  N  I  E  R  E  E
U  O  C  L  D  H  L  R  A  T  M  T
N  N  R  L  N  I  Y  U  E  E  O  I
I  D  M  A  U  M  Q  M  T  A  L  L
M  A  R  G  O  L  I  K  P  S  I  I
A  T  N  I  P  T  E  L  O  P  K  L
R  G  A  L  N  P  U  C  E  O  E  L
G  E  R  E  I  L  C  T  O  O  F  I
T  E  C  N  U  O  N  D  U  N  R  M
```

GA1504

Backwards, Forwards

Palindromes are words that are spelled the same forwards and backwards. Here are some examples: noon, bib, radar.

Here are some examples of number palindromes: 11, 22, 606, 9229. The first 3-digit palindrome number is 101.

a. Can you find the next four 3-digit palindrome numbers?

   Write them here: _____ _____ _____ _____

b. What would the largest 3-digit palindrome number be?_____

c. The year 1881 is a palindrome. What was the most recent past year that was a palindrome? _____

d. What is the next year in the future that will be a palindrome? _____

e. Make up your own palindrome numbers that have:

   1. 7 digits _____

   2. 8 digits and only even numbers _____

   3. 10 digits and only odd numbers _____

Number Shapes

Here are some *square* numbers:

a. Figure out the next square number and draw it here.

b. Predict what the next square number will be. _____

Here are some *triangle* numbers.

c. What is the next triangle number? _____ Draw it here.

d. Predict what the next triangle number will be. _____

Continue making square and triangle numbers until you see the patterns that work. Can you explain each pattern?

e. The pattern for square numbers:

f. The pattern for triangle numbers:

GA1504

Slick Picks

Four toothpicks have been used to form this square:

Notice that it takes only 7 toothpicks to make 2 squares.

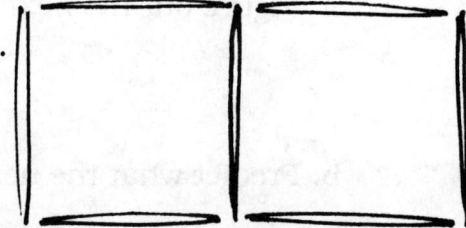

1. Now draw 3 connected squares made with toothpicks. How many toothpicks are needed?

2. Now add a 4th connected square. How many toothpicks are used?

3. Can you describe a pattern that seems to work for adding squares in a line?

4. There's a way to make 4 connected squares using just 12 toothpicks. Can you find it? Draw your answer here.

5. **Using 12 toothpicks, see if you can use them all to make arrangements containing 1 square, 2 squares, and 3 squares. Which are possible? Which are not?

94

Counting Confusion

How many triangles can you find? _____

How many squares can you find? _____

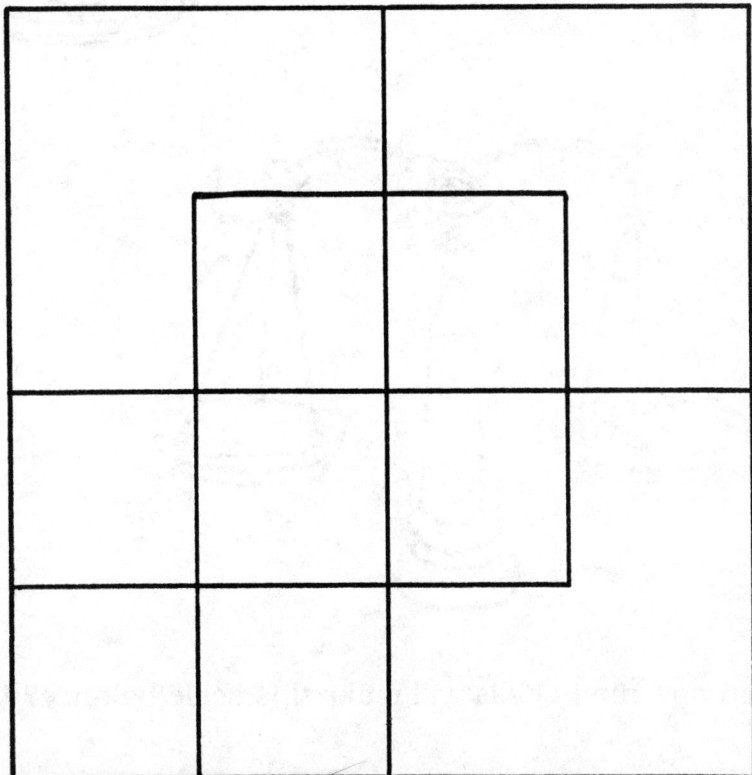

A Weighty Problem

If

and

then how many balls will make this scale balance? Draw your answer.

Find the Fact

Study each picture. Circle the one sentence in each set that is true.

1. There are six shapes in all.
2. There are less balloons than hats.
3. There are less hats than balloons.

4. Two toys are less than 16¢.
5. All the toys are more than 17¢.
6. All the toys are less than 20¢.

7. All are rectangles.
8. Three have 4 sides.
9. Two are triangles.

10. All have the same sum.
11. All have different sums.
12. The first sum is more than the last sum.

$$\begin{array}{r} 3 \\ +5 \\ \hline \end{array} \qquad \begin{array}{r} 4 \\ +4 \\ \hline \end{array} \qquad \begin{array}{r} 6 \\ +2 \\ \hline \end{array}$$

GA1504

Zeeps

All of these are Zeeps:

None of these are Zeeps:

Which of these are Zeeps? Circle your answers.

A.

B.

C.

D.

E.

F.

98

Googles

All of these are Googles:

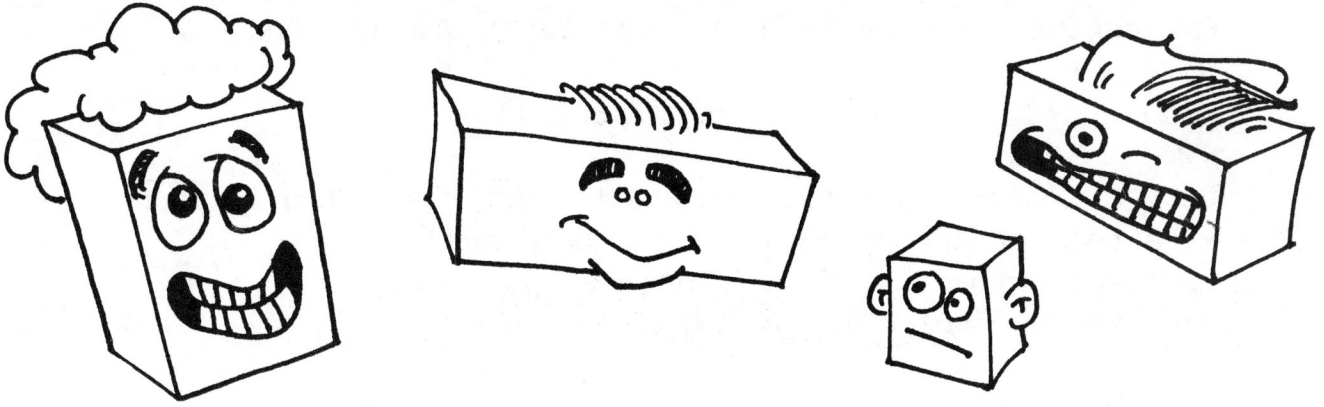

None of these are Googles:

Which of these are Googles? Circle your answers.

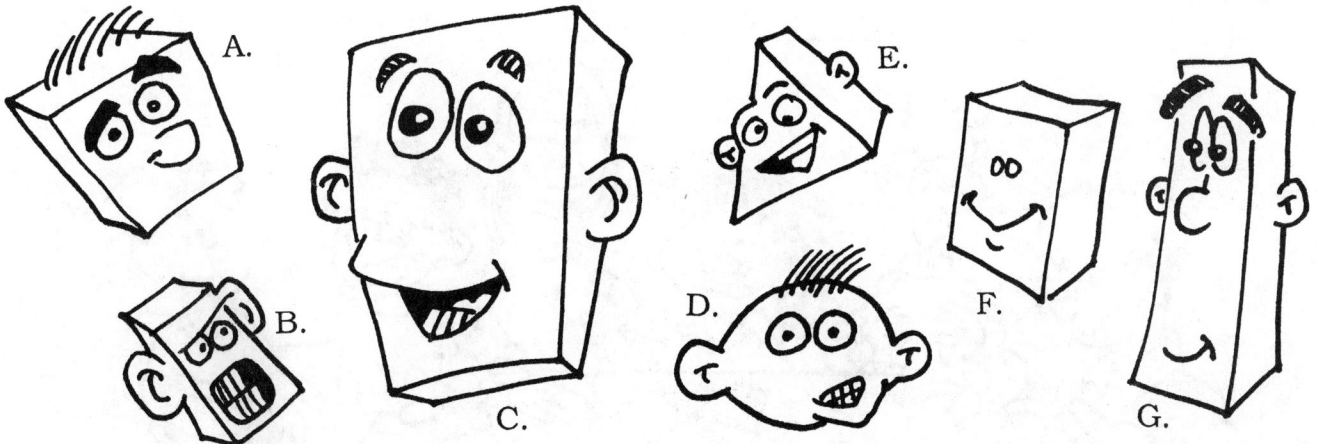

A.

B.

C.

D.

E.

F.

G.

GA1504

What's My Number?

A. Mr. Bowen's class is playing a number game. Three of his students–Cathy, Pat, and Drew–have each secretly chosen one of these numbers:

3, 5, 6, 7, 8, 9, 11

Using these clues, try to figure out which student chose which number.
1. Pat chose an odd number; the others chose even numbers.
2. Cathy's number is larger than Andrew's.
3. Pat's number is one less than Andrew's number.

Write your answers here.

Cathy _____ Pat _____ Andrew _____

B. Now three more students–Lynne, Sarah, and David–are playing the game. Here are the numbers they chose from.

11, 12, 14, 17, 18, 20, 23

Use these clues to tell who chose which number.
1. David's number is less than 20.
2. No one chose an odd number.
3. Sarah's number is 6 less than David's number.
4. Lynne's number is larger than David's number.

Write your answers here.

Lynne _____ Sarah _____ David _____

Group Troop

These belong to the same group:	These do *not* belong:	Circle which of these belong:
a. 6, 82, 14, 20	19, 33, 7, 21	30, 9, 51, 46, 8
b. eighty, twelve, ninety, twenty	five, ten, forty, thirteen	thirty, three, six, eleven, nine
c. cup, gallon, liter, teaspoon	kilogram, yard, ounce, meter	milliliter, pound, mile, pint
d. 30, 45, 15, 50	23, 14, 61, 42	25, 39, 21, 60, 75
e. July, March, May, October	September, April, February	June, August, November, January
f. pyramid, box, sphere	oval, square, rectangle	cylinder, circle, triangle, cube

**Now make up a rule for your own group. Fill out all three columns in this chart for your group. See if a classmate can circle the correct answers.

Answer Key

King Key, Page 2
29 Keys

Locker Duty, Page 3
You will paint the number 2 fourteen times.

Leap Frog, Page 4
c. Frank needs 6 jumps for 12 spaces.
d. Franny needs 5 jumps for 15 spaces.

Three's the Key, Page 5

100	86	85	86	87	91	94	99	100
98	82	83	84	89	90	92	95	97
80	79	81	79	16	14	93	20	19
77	78	3	6	9	12	15	18	22
76	75	5	8	13	11	19	21	20
68	72	70	52	49	47	26	24	26
65	69	67	51	48	45	46	27	31
64	66	61	54	50	42	41	30	34
62	63	60	57	58	39	36	33	40

Counting Scramble, Page 7
Unscrambled Words:
nine   four
eight   six
second   fourth
ten   five
seven   twenty
one   thirteen
third

```
        S E C O N D
          F O U R
        F O U R T H
          N I N E
            T E N
          F I V E
      S E V E N
        E I G H T
        T W E N T Y
          O N E
        T H I R T E E N
T H I R D
            S I X
```

Animal Antics, Page 8
A WALKIE-TALKIE

Ann's Antiques, Page 9
thimble–2, toaster–9, camera–7, safety pin–5, saxophone–4, scissors–3, piano–8, microwave oven–10, glasses–1, stapler–6

What Fits? Page 10
Box A–Even numbers, Box B–Odd numbers, Box C–Numbers less than 10

Number Detective, Page 11
Numbers with less than 5 ones–14, 21, 23
Numbers with 2 tens–28, 26, 25
Numbers greater than 30–36, 34, 31

Circular Sums, Page 16

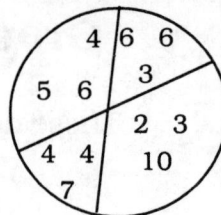

Problem Search, Page 17

10	-	7	=	3	+	6	=	9
+		+		+		+		-
8	+	1	=	9	-	4	=	5
=		=		=		=		=
18		8		12		10		4
-		+						+
9	+	7	=	16	-	8	=	8
=		=						=
9		15	-	9	=	6		12

Sum Squares, Page 18

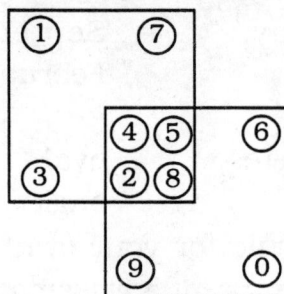

Pick-a-Path, Page 19
1. You can reach a total of 49 by going through every number in this order: 4-6-10-7-3-2-8-9. Variations are possible.
2. The lowest total is 12; go through 7-3-2.

Bull's-Eye I, Page 20
Greg–25, Danny–28, Jan–27, Kate–27, David–23

Bull's-Eye II, Page 21
(Other combinations may be possible.)
Greg: 3-10's, 2-1's   Danny: 3-8's, 2-3's
Jan: 4-8's, 1-3   Kate: 2-10's, 2-8's, 1-3
David: 2-10's, 3-8's

Pet Store Prices, Pages 22-23
mouse–73¢, rabbit–52¢, kitten–79¢, crab–24¢, snake–50¢, lizard–70¢, puppy–94¢, iguana–53¢, parrot–88¢
One animal that would cost more than one dollar is the vulture ($1.19). Other answers are also possible.

GA1504

Domino Dilemma, Page 24
1. 1 + 1, 0 + 2, 0 + 3, 1 + 2
2. 11 total dots
3. 6 + 6, 6 + 5, 5 + 5, 6 + 4, 5 + 4, 6 + 3
4. 61 dots
5. 97 dots
6. 133 dots

Book Drops, Page 25
Van A-25, Van B-41, Van C-34

Beanbags, Page 26
These numbers should be circled: 65, 80, 100, 49, 60, 85, 68, 47, 46, 50, 56
For a score of 84, 2 beanbags would land in the 17 spot, and 2 in the 25 spot.

Bingo! Page 27
A = 9, B = 16, C = 55, D = 62

Test Drive, Pages 28-29
Total miles driven: Clint–669, Michelle–896, Ken–587, Elizabeth–492, Robert–972
Routes #1 and #2: Wyatt–South Bend–Knox
Wyatt–Lawrence–Mendon
Route #3: Hartford–Lawrence–Niles–Knox
(Each route could also be reversed.)

Quilt Crafts, Page 30
1. 12
2. 4
3. 8

Carnival Candy, Page 31
A. 4 pieces, 8¢   C. Sally spent more.
B. 3 pieces, 9¢   D. 5 pieces

Potato Problem, Page 33
MASH THEM FIRST

Dice Doodles, Page 35
These are the possible combinations:
2 x 6   3 x 4   3 x 5   3 x 6   4 x 4
4 x 5   4 x 6   5 x 5   5 x 6   6 x 6

A-Maze-ing 40, Page 36

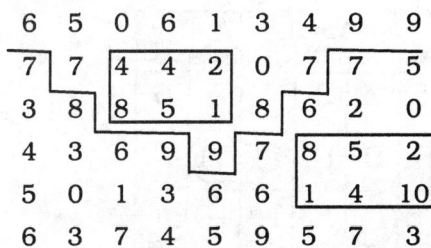

```
6 5 0 6 1 3 4 9 9
7 7 4 4 2 0 7 7 5
3 8 8 5 1 8 6 2 0
4 3 6 9 9 7 8 5 2
5 0 1 3 6 6 1 4 10
6 3 7 4 5 9 5 7 3
```

Cross Out, Page 37
Message: You're GREAT!

Missing Signs, Page 38
1. 4 x 4 = 16
2. 3 = 21 ÷ 7
3. 20 ÷ 10 = 2
4. 8 x 3 = 24
5. (5 - 4) x 2 = 2
6. (6 ÷ 3) x 5 = 10
7. (12 ÷ 3) + 4 = 8
8. (10 - 4) + 6 = 12
9. (16 ÷ 8) - 2 = 0
10. (15 + 10) - 5 = 20

Bingo 48, Page 39
The fourth horizontal row is the BINGO.

What's Left? Page 40
Here is one possible solution:

	Remainder 1	Remainder 2	Remainder 3	Remainder 4	Remainder 5
Divide by 3	13	17	X	X	X
Divide by 4	21	18	7	X	X
Divide by 5	26	22	23	34	X
Divide by 6	49	8	15	28	41

Domino Dare I, Page 42

```
A.    365        B.    604
     x  4             x  6
     1460             3624
```

Domino Dare II, Page 43

```
A.    123        B.    452
     x  9             x  6
     1107             2712
```

Rewarding Work, Page 44
Bob worked the most hours.
Bob: 2,016; Shari: 1,932; Mary: 1,887; Randy: 1,900
**The engraving will cost $11.20.

Table Time, Page 45

x	17	23	49	78
56	952	1,288	2,744	4,368
90	1,530	2,070	4,410	7,020
34	578	782	1,666	2,652
62	1,054	1,426	3,038	4,836

GA1504

Four Farmers, Page 47

D	J		P	R

Fractional Sports, Page 49
1. Hockey
2. Archery
3. Boxing
4. Skiing
5. 3/5, 2/5, 2/5
6. 3/5, 1/4, 1/2

Star Search, Page 50

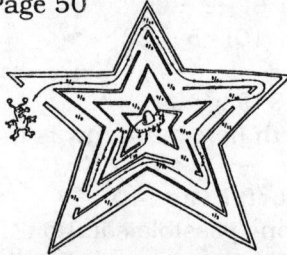

Shady Fractions, Page 51

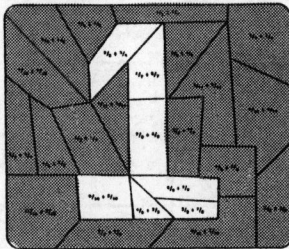

Pocket Change, Page 52
With 12 coins, Ryan would have $1.26.
With 24 coins, Ryan would have $2.52.
With 36 coins, Ryan would have $3.78.

Laugh a Little! Page 54
a. I. 14, II. 17, III. 28, IV. 33, V. 7
b. Chapter IV is longest; Chapter V is shortest.
c.

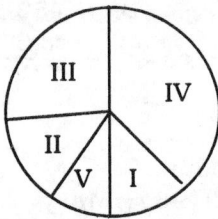

Correct the Collections, Page 55

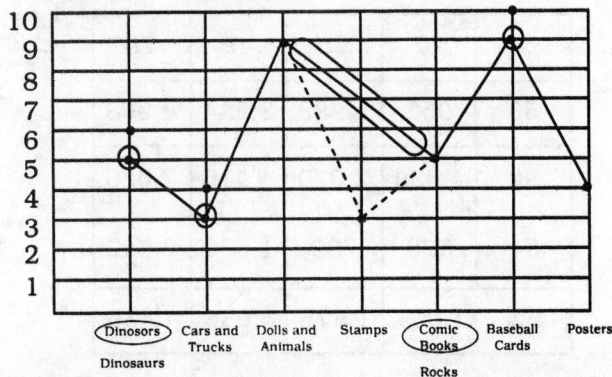

Graph a Giraffe! Page 57

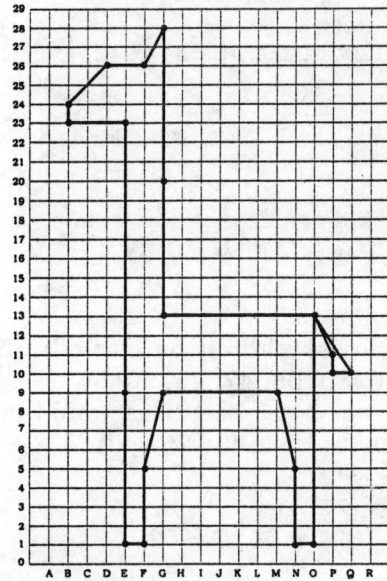

What Is It? Page 59

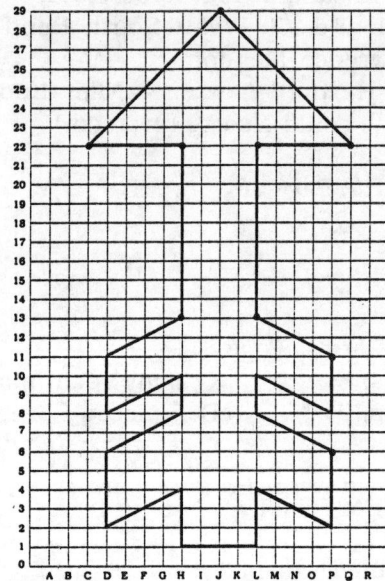

What's Wrong? Page 60

GA1504

Becky's Birthday, Page 61
Answer: March 22

Calendar Computations, Page 62
1. 30
2. 30 + 31 = 61
3. 61 x 7 = 427
4. 427 x 4 = 1708
5. 1708 + 366 = 2074
6. 2074 - 12 = 2062
Answer: the year 2062

What's Next? Page 63
Message: You are right!

Watch Match, Page 64
1. E   5. D
2. F   6. A
3. B   7. C
4. extra

Taxi Time, Page 65
8, 4, 11, 2, 6, 12, 9, 1, 5, 10, 7, 3

Clock Calamity, Page 66
Missing numeral = XI (11)

More or Less, Page 67
1. IV   6. XX
2. VI   7. LX
3. XV   8. IX
4. XL   9. XIX
5. XI   10. LXX
Largest: LXX

Shape Up! Page 68
Picture D

Tricky Triangles, Page 70
All six triangles will fit in if they are arranged like this:

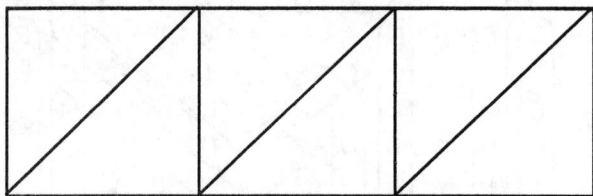

Common Couples, Page 71

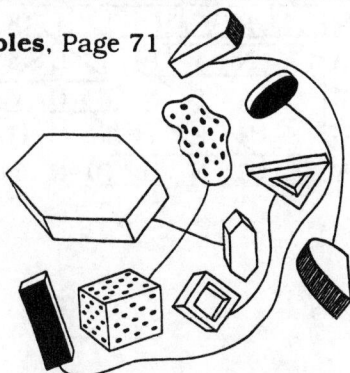

Shape Mistake, Page 72
A. The 4th shape
B. The last shape
C. The second shape
D. The third shape
E. The first shape

Folding Fits, Page 73

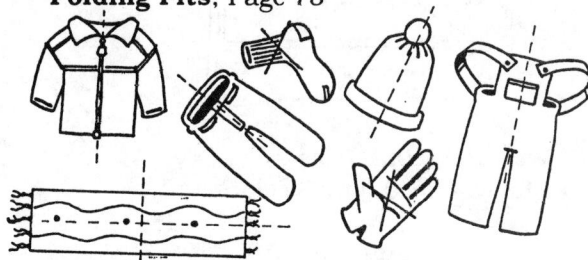

Baffling Blocks, Page 74
Block A can go 2 different ways.
Block B can go 3 different ways.
Block C can go 4 different ways.

Square Puzzlers, Page 75

A.   B.   C.

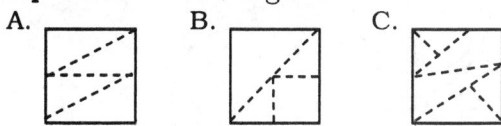

A Neat Drawer, Page 76

Wall Painting, Page 77
She will need only 3 colors.

Floor Samples, Page 78

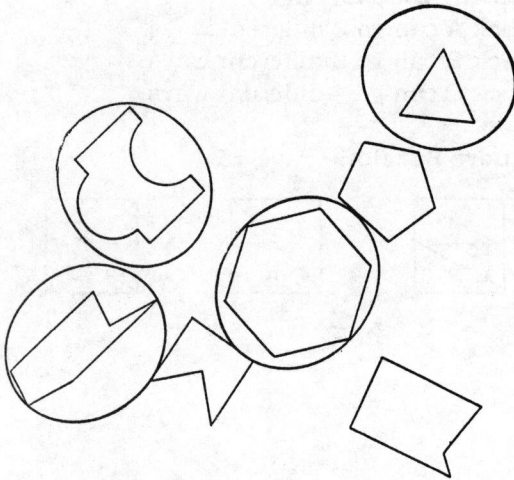

Shape Caper, Page 79
Shapes B and E

Rolling Coins, Page 82
A. Nathan–43¢   Justin–53¢
B. Move a nickel from Justin's pile to Nathan's pile.

Barney's Banking, Page 83
1. 3 nickels
2. 1 dime, 5 pennies
3. 4 nickels
4. 2 dimes, 5 pennies
5. 1 quarter, 1 penny
6. 1 quarter, 1 nickel, 1 penny
7. 1 quarter, 1 dime, 1 nickel, 2 pennies
8. 2 quarters, 1 nickel
9. 6 dimes, 2 nickels
10. 3 quarters, 2 dimes

,Barney's Bill, Page 84
1. 3-$20s, 2-$10s, 2-$5s
2. 1-$10, 4-$5s
3. 8-$5s, 5-$1s
4. 1-$20, 6-$10s, 1-$5
5. 2-$20s, 2-$5's, 10-$1's
6. 1-$20, 4-$10's, 2-$5s
7. 3-$20s, 15-$1s
8. 4-$20s, 1-$10, 2-$5s
9. 2-$20s, 5-$10s, 2-$5s
10. 3-$20s, 3-$10s, 10-$1s
**one
***10-$5s

Tools of the Trade, Page 85
Hacksaw, wrench, hammer, drill

Mabel's Labels, Page 87
Potatoes, apples, laundry soap, ham, salt–pounds; juice, granola bars, pepper, soup, cereal–ounces

Temperature Travels, Page 88
A. New York City
B. Tucson
C. Miami
D. Chicago

Measurement Mess, Page 89

man–g	quarter–i	juice–a
turtle–l	road–k	oil–b
elephant–e	snake–d	milk–h
apple–f	window–j	water–c

Tile Style, Page 90
2. 11 cm², 3. 8 cm², 4. 16 cm², 5. 9 cm², 6. 8 cm², 7. & 8. Outcomes may vary.

Abbreviated Word Search, Page 91

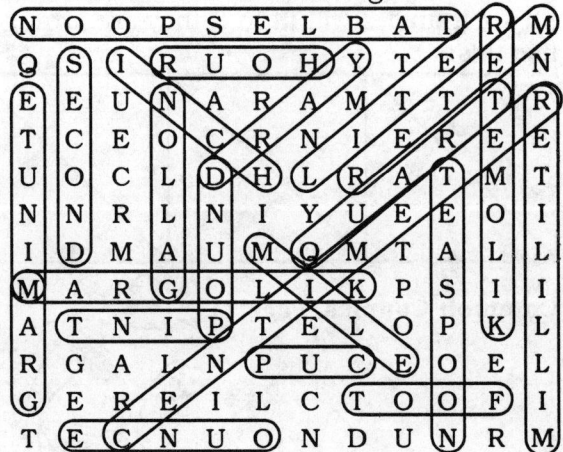

GA1504

Backwards, Forwards, Page 92
a. 111, 121, 131, 141
b. 999
c. 1991
d. 2002
e. Answers may vary. Here are some possibilities:
1. 1,234,321
2. 24,688,642
3. 1,357,997,531

Number Shapes, Page 93
a. 25
b. 36
c. 15
d. 21
e. Multiply a number by itself. Teacher: This is a good way to introduce exponents.
f. Add the next whole number to the last triangle number. 1 + $\underline{2}$ = 3, 3 + $\underline{3}$ = 6, 6 + $\underline{4}$ = 10, 10 + $\underline{5}$ = 154

Slick Picks, Page 94
1. 10
2. 13
3. Add 3 toothpicks for each additional square.
4. 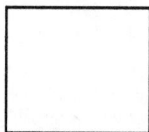 (Actually, you've just made $\underline{5}$ squares–the large outer square is the fifth one!)

Here is another solution discovered by a student:

5. 1 square:

3 squares:

2 squares may not be possible.

Counting Confusion, Page 95
Triangles: 10
Squares: 13

A Weighty Problem, Page 96
You should draw 4 balls.

Find the Fact, Page 97
3, 6, 8, 10

Zeeps, Page 98
A, C, and E

Googles, Page 99
B, C, and F

What's My Numbers?, Page 100
A. Cathy–8, Pat–5, Andrew–6
B. Lynne–20, Sarah–12, David–18

Group Troop, Page 101
A. 30, 46, 8 (even numbers)
B. thirty, eleven (six-letter number words)
C. milliliter, pint (units that measure volume)
D. 25, 60, 75 (multiples of 5)
E. August, January (months with 31 days)
F. cylinder, cube (3-dimensional figures)

GA1504